Professional BlackBerry

Professional BlackBerry®

Craig James Johnston with Richard Evers

Wiley Publishing, Inc.

Professional BlackBerry®

Published by
Wiley Publishing, Inc.
10475 Crosspoint Boulevard
Indianapolis, IN 46256
www.wiley.com

Published simultaneously in Canada

ISBN-13: 978-0-7645-8953-9
ISBN-10: 0-7645-8953-9

Manufactured in the United States of America

10 9 8 7 6 5 4 3 2 1

1B/RT/QX/QV/IN

For general information on our other products and services or to obtain technical support, please contact our Customer Care Department within the U.S. at (800) 762-2974, outside the U.S. at (317) 572-3993 or fax (317) 572-4002.

Wiley also publishes its books in a variety of electronic formats. Some content that appears in print may not be available in electronic books.

Library of Congress Cataloging-in-Publication Data

Johnston, Craig J., 1967-
  Professional BlackBerry / Craig J. Johnston with Richard Evers.
     p. cm.
  Includes bibliographical references and index.
  ISBN-13: 978-0-7645-8953-9 (paper/website)
  ISBN-10: 0-7645-8953-9 (paper/website)

  1. BlackBerry (Computer) 2. Pocket computers. 3. Mobile computing. I. Evers, Richard. II. Title.
  QA76.8.B53J56 2005
  004.16—dc22
                                        2005011805

This book is a publication of John Wiley & Sons. Research In Motion Limited (RIM) neither endorses the book nor accepts any responsibility whatsoever for the content or presentation of the materials contained within it.

*To my loving wife, Karen Fayd'herbe de Maudave Johnston.
I'm enjoying our journey together. —CJJ*

*To my wife, Donna, and our children, Stephanie, Samantha,
and Adam. —RE*

About the Authors

Craig James Johnston of East Windsor, New Jersey, has more than 15 years of networking experience, most recently with the BlackBerry. He has done proof-of-concept BlackBerry projects and has actively supported BlackBerry devices in a Lotus Domino environment since 2000. His extensive knowledge of networking, hardware, and wireless technologies is coupled with writing and technical instruction.

Richard Evers of Waterloo, Ontario, Canada, is the editor of the *BlackBerry Developer Journal*. He is an expert in the areas of wireless communication and small-footprint application development. He has more than 25 years' experience designing and developing commercial and custom applications. He has been editor and publisher of numerous publications, including *Transactor* magazine. He creates and publishes educational Web sites, and he develops customized Web software (including search engines, custom proxy servers, and browsers).

Credits

Acquisitions Editor
Debra Williams Cauley

Development Editor
Kevin Shafer

Technical Editor
Richard Evers

Copy Editor
Michael Koch

Editorial Manager
Mary Beth Wakefield

Vice President and Executive Group Publisher
Richard Swadley

Vice President and Publisher
Joseph B. Wikert

Project Coordinator
Erin Smith

Graphics and Production Specialists
Jonelle Burns
April Farling
Lauren Goddard
Denny Hager
Jennifer Heleine
Melanee Prendergast
Amanda Spagnuolo

Quality Control Technicians
Leeann Harney
Jessica Kramer
Carl William Pierce
Brian Walls

Proofreading and Indexing
TECHBOOKS Production Services

Contents

Contents

Contents

Contents

Contents

Acknowledgments

While writing this book, I had the assistance of a few people at RIM who (unofficially) gave up their personal time to review parts of my original manuscript to make sure that it is as accurate as possible and that it provides the most value to you, the reader. These people include Michael Clewley, Jennifer Emery, Peter Hantzakos, Warren James, Vaibhav Joshi, Brad Maybee, Thomas Pfeifer, Sassan Sanei, Mark Sohm, and Roger Werner.

I also thank my wife, Karen Johnston, who put up with not having me around during the many months it took to write this book.

Introduction

The RIM BlackBerry has become synonymous with always-on, real-time wireless e-mail. Other companies provide similar solutions, but they are a very distant second to the wide acceptance of the BlackBerry. BlackBerry usage is virtually exploding right before our eyes, and in some cities around the world, you see a BlackBerry user every few steps. There is even a new affliction that has been termed "BlackBerry Thumb," which is a repetitive stress injury to the thumbs from using them in ways in which they were never meant to be used.

The BlackBerry can be used for real-time push e-mail, but it also supports real-time wireless synchronization of Personal Information Manager (PIM) data and has the capability to connect the user to systems back in the office in a very secure manner. The aim of this book is to cover two main topics that relate to the BlackBerry environment. In the first section, we start off by discussing how the BlackBerry works and the different components that make up the BlackBerry environment. We then move on to discussing how to install a BlackBerry server and how to prepare for a BlackBerry pilot. Finally, we discuss how to roll out BlackBerrys and the desktop software (if you need it) and how to maintain your BlackBerry environment.

In the second section we discuss the different ways in which you can extend the functionality of the BlackBerry beyond just email and PIM synchronization. In this section, we discuss Mobile Data Service (MDS), the BlackBerry Channel, Web portals, handheld Java applications, and a unique mobile development language called Plazmic. All of these technologies are extremely powerful to the developer and end user. MDS is the core of all data transfer between the handhelds and the corporate network, and we discuss the many different ways that it can be used by developing your own applications or using off-the-shelf solutions.

Many people are not aware that every BlackBerry running version 3.7 or later of the Handheld Software includes something called the Plazmic Media Engine. This incredible engine enables the BlackBerry to display stimulating rich Web content. We show you how to use the Plazmic Developer Kit to create these visually stimulating Web environments while keeping the size of the content small.

Being in mobile computing for a while now, largely supporting BlackBerry devices, I felt that someone needed to write a book that dealt with support and development of BlackBerry devices. Companies that already have a BlackBerry infrastructure seem reluctant to develop the platform further, even though it requires very little effort to produce a large result. This large result always means that your user community benefits and, in most cases, allows your company to make more money. This is because the BlackBerry users can do more while outside the office. This ultimately makes them much more productive and happier that they do not have to come into the office as often.

If you already have BlackBerry devices and would like to extend their functionality, or you would like to create a proof of concept portal or channel to demonstrate to your organization that it can easily be done, then you will find exactly what you are looking for here. Companies that do not have BlackBerry devices will find this book useful in understanding how the BlackBerry works and how it may fit into their organizations.

We discuss the latest version of the BlackBerry, version 4.0, but will also cover the older Exchange 3.6 and Domino 2.2 versions.

Throughout this book, any time I needed to show a screenshot of the BES 4.0 Management Console, I've used the Lotus Domino BlackBerry Manager. I did this because the visuals are more appealing and look more self-explanatory on the screen than the Exchange 4.0 BES BlackBerry Manager. Even so, if you use the Exchange BES 4.0 BlackBerry Manager, you will find that the menus are in the same structure in most cases, and, where they may differ, it is very easy to figure out where to find them.

Whom This Book Is For

This book is intended primarily for two kinds of readers. The first group consists of IT staff who will be supporting and maintaining the BlackBerry environment within their companies. They should already be familiar with their e-mail system and the concepts that surround them. The second group includes developers who are familiar with HTML, Wireless Markup Language (WML), and Java. These would be people who would do internal development within an organization.

Although the book is intended primarily for these kinds of readers, we provide enough information in each chapter to allow all technically savvy readers to follow along and understand the concepts.

If you will be supporting primarily your BlackBerry environment and not doing any development, Chapters 1 through 8 are for you. If you are interested only in extending the functionality of your BlackBerry environment or providing proof of concept that shows how it can be done, then Chapters 9 through 14 are for you. Chapters 9 through 14 do not rely on the previous chapters in any way, so you can skip right to Chapter 9.

What This Book Covers

We cover the technologies that make up the BlackBerry infrastructure, including components that run within your organization, components that run at Research In Motion (RIM), the cellular carrier, the cellular data networks, and the handhelds themselves. We discuss BlackBerry Enterprise Server 4.0 and Handheld Software 4.0, but we don't forget about the older versions and so we cover Exchange BlackBerry Enterprise Server 3.6 and Domino BlackBerry Enterprise Server 2.2.

How This Book Is Structured

The book is divided into two parts. The first part covers supporting the BlackBerry. In Chapters 1 through 8 you will find such topics as how the components of the BlackBerry infrastructure work, how to pilot BlackBerrys within your organization, how to roll out BlackBerry devices to your users, and how to plan disaster-recovery scenarios to provide the maximum uptime for your users.

The second part of the book (Chapters 9 through 14) covers the development side of the BlackBerry. In these chapters, you will find details and sample scripts that will enable you to take the ideas learned in this book and apply them to your BlackBerry environment. These include how to create a BlackBerry intranet portal, how to create a BlackBerry channel, how to write handheld-side J2ME code, and how to spice up your intranet portal with Plazmic.

Appendixes A through F provide useful reference information to help you understand and expand upon Chapters 9–14.

What You Need to Use This Book

If you will be reading Part II of this book, you will need the BlackBerry 4.0 Java Development Environment (JDE). This tool comprises the JDE plus a BlackBerry simulator, which will allow you to see all of our sample code running on the simulated BlackBerry. You can download the JDE from www.blackberry.com/developers/na/java/index.shtml.

Conventions

To help you get the most from the text and keep track of what's happening, we've used a number of conventions throughout the book.

Tips, hints, tricks, and asides to the current discussion are offset and placed in italics like this.

As for styles in the text:

❑ We *highlight* new terms and important words when we introduce them.

❑ We show keyboard strokes like this: Ctrl+A.

❑ We show file names, URLs, and code within the text like so: `persistence.properties`.

❑ We present code in two different ways:

```
In code examples we highlight new and important code with a gray background.
```

```
The gray highlighting is not used for code that's less important in the present
context, or has been shown before.
```

Source Code

As you work through the examples in this book, you may choose either to type in all the code manually or to use the source code files that accompany the book. All of the source code used in this book is available for download at www.wrox.com. When at the site, simply locate the book's title (either by using the Search box or by using one of the title lists) and click the Download Code link on the book's detail page to obtain all the source code for the book.

Because many books have similar titles, you may find it easiest to search by ISBN; this book's ISBN is 0-7645-8953-9.

After you have downloaded the code, just decompress it with your favorite compression tool. Alternately, you can go to the main Wrox code download page at www.wrox.com/dynamic/books/download.aspx to see the code available for this book (and all other Wrox books).

Errata

We make every effort to ensure that there are no errors in the text or in the code. However, no one is perfect, and mistakes do occur. If you find an error in one of our books, like a spelling mistake or faulty piece of code, we would be very grateful for your feedback. By sending in errata you may save another reader hours of frustration and at the same time you will be helping us provide even higher quality information.

To find the errata page for this book, go to www.wrox.com and locate the title using the Search box or one of the title lists. Then, on the book details page, click the Book Errata link. On this page you can view all errata that has been submitted for this book and posted by Wrox editors. A complete book list including links to each's book's errata is also available at www.wrox.com/misc-pages/booklist.shtml.

If you don't spot "your" error on the Book Errata page, go to www.wrox.com/contact/techsupport.shtml and complete the form to send us the error you have found. We'll check the information and, if appropriate, post a message to the book's errata page and fix the problem in subsequent editions of the book.

p2p.wrox.com

For author and peer discussion, join the P2P forums at p2p.wrox.com. The forums are a Web-based system for you to post messages relating to Wrox books and related technologies and interact with other readers and technology users. The forums offer a subscription feature to e-mail you topics of interest of your choosing when new posts are made to the forums. Wrox authors, editors, other industry experts, and your fellow readers are present on these forums.

At http://p2p.wrox.com you will find a number of different forums that will help you not only as you read this book, but also as you develop your own applications. To join the forums, just follow these steps:

1. Go to p2p.wrox.com and click the Register link.

2. Read the terms of use and click Agree.

3. Complete the required information to join as well as any optional information you want to provide and click Submit.

4. You will receive an e-mail with information describing how to verify your account and complete the joining process.

   You can read messages in the forums without joining P2P but in order to post your own messages, you must join.

After you've joined, you can post new messages and respond to messages other users post. You can read messages at any time on the Web. If you would like to have new messages from a particular forum e-mailed to you, click the Subscribe to this Forum icon by the forum name in the forum listing.

For more information about how to use the Wrox P2P, be sure to read the P2P FAQs for answers to questions about how the forum software works as well as many common questions specific to P2P and Wrox books. To read the FAQs, click the FAQ link on any P2P page.

Professional BlackBerry®

Part I

Understanding, Planning for, and Installing BlackBerry

System Architecture

It is always important to understand the architecture of any product that you must support or on which you may develop. Understanding the different components that make up the Research In Motion (RIM) BlackBerry platform will help you visualize how it integrates with your e-mail system, how the handheld units communicate with your network, and where to place the different components that make up the BlackBerry infrastructure so that they maximize performance and security.

We will start by discussing the latest version of the BlackBerry platform, version 4.0, which shipped in early December 2004. This new version offers significant improvements in performance, management, user features, and application development. Because not everyone will upgrade immediately to version 4.0, we will also discuss the previous version of the BlackBerry platform, which is version 3.6 for Microsoft Exchange and version 2.2 for Lotus Domino.

Current Architecture (Version 4.0)

BlackBerry 4.0 represents a big leap forward in the BlackBerry platform technology. It offers substantial benefits to organizations in many areas, including operations, development, user features, handheld software improvements, BlackBerry administration, and internal network design.

In Figure 1-1, you can see the current BlackBerry 4.0 architecture. It shows how all of the BlackBerry components remain behind the corporate firewall with the mail servers and application servers while communicating with the handheld units through the firewall, via the BlackBerry infrastructure. You will also notice two paths from the handheld units to the BlackBerry components. Path 1 is from the BlackBerry components behind the firewall, through the BlackBerry infrastructure to the handheld units on the wireless networks. Path 2 is through the local area network (LAN) to the handheld unit connected to a USB or serial port on a computer. We will discuss these different components and the methods of communication in more detail later in this chapter.

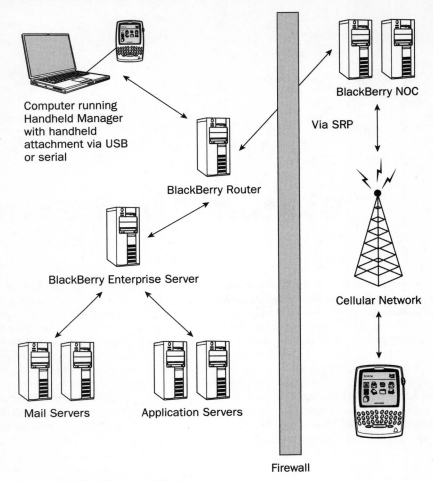

Figure 1-1: BlackBerry architecture

This section describes the basic components included in BlackBerry 4.0:

- ❑ Handheld unit
- ❑ BlackBerry Enterprise Server (BES)
- ❑ BlackBerry Network Operations Center (NOC)
- ❑ Data network
- ❑ Attachment Service
- ❑ Mobile Data Service (MDS)

❑   BlackBerry Router

❑   Configuration Database

Handheld Unit

The name *BlackBerry* can be traced back to Lexicon, a company that had been contracted in 1998 by RIM to develop a name for RIM's new wireless e-mail device. The company presented RIM with approximately 75 name candidates, but the word that immediately stood out was *blackberry*. The name has perfect symmetry in that *black* and *berry* have five letters each. RIM felt that words that ended in *y* were approachable. They liked the playfulness of the name, and the color scheme fit well with the color of the device. In those beginning days, the handheld devices connected to two data-only networks run by Bell South and Motient, which operated in the United States and Canada. The Bell South network was called the Mobitex network. Bell South later merged with Cingular and so the network fell under the Cingular flag. The Mobitex network is still running and there are still data-only BlackBerrys available, such as the RIM 950, RIM 957, and the BlackBerry 5790, that connect to it. Motient ran the other network, called the DataTAC network. It still runs today in the USA and the RIM 850 and RIM 857 connect to it. In December 2004 the DataTAC network was shut down in Canada.

The data-only BlackBerry handheld units have no telephone. Because they do not run on a cellular network, they can be used only within the United States and Canada. These handheld units still have a place in today's always-connected world. They can be used in many situations where the company doesn't need the employee to use the telephone, and the employee does not travel outside North America. These situations include employees who travel extensively, but must stay in touch through e-mail. These same employees may need to use the BlackBerry to keep schedules current and they may need to use custom applications to perform data-entry tasks on job sites or to fill out work reports.

One final point about these older handheld units (such as the RIM 850, RIM 857, RIM 950, and RIM 957) is that they run on an operating system that is built on the C++ programming language. This means that any development for these handheld devices must be done using C++. The Web browser on these handheld units can interpret only Wireless Markup Language (WML) and cannot process Hypertext Markup Language (HTML), Extensible Hypertext Markup Language (XHTML), or JavaScript. One exception to this rule is the new BlackBerry 5790, which is a data-only BlackBerry that runs on the Cingular Mobitex network, but is a Java device. Because of this, it shares its feature set with all of the other Java handheld units (which we will discuss next), with the exception of having no phone.

The next generation of BlackBerry handheld devices was the Java handheld units. These handheld devices run on an operating system built entirely in Java. This means that all development for these handheld units must be done using Java 2 Mobile Edition (J2ME). The Web browser on these handheld units can process WML and WMLScript if operated through a WAP gateway. (Appendix B discusses WMLScript in more detail.) If operating through a BlackBerry Enterprise Server gateway, or the BlackBerry Internet Browsing Service (BIBS) gateway, the BlackBerry Browser can also process HTML, cHTML, and XHTML pages. It can also process JavaScript and use style sheets in limited form if the handheld unit is running the BlackBerry version 3.8 or 4.0 software.

The first J2ME BlackBerry handheld units were built to connect to the Global System for Mobile Communications (GSM) cellular voice network and the data network that runs on top of GSM, General Packet Radio Service (GPRS). After a few models of the GSM/GPRS BlackBerry were released, RIM designed J2ME BlackBerrys that ran on other cellular technologies such as Motorola's Integrated Digital Enhanced Network (iDEN) and Code Division Multiple Access (CDMA).

All of the C++ handheld devices have monochrome displays, while the J2ME handheld units feature either monochrome or color displays. The handheld devices with monochrome displays have a screen resolution of 160 × 160 pixels, while the color devices range in screen resolution from 160 × 240 pixels to 240 × 260 pixels.

The actual physical size of the screen may not always convey the screen resolution or bit density. For example, the BlackBerry 5790 and 6710/6720 models have large screens, but the actual bit density is 160 × 160 pixels. The BlackBerry 7700 series has the same physical screen dimensions, but the bit density is 240 × 240 pixels. The BlackBerry 7100 series has an even more tightly packed bit density, squeezing 240 × 260 pixels into a physical screen that is actually smaller than the older BlackBerrys.

Besides BlackBerry handheld devices made by RIM, you can purchase handheld units from other manufacturers that have the BlackBerry Connect or BlackBerry Built-In technology. *BlackBerry Connect* is software that can be loaded onto a cellular telephone or other wireless device that allows it to be used as a BlackBerry. The native e-mail, calendar, and address book applications on the handheld units are used, but the transport mechanism is provided by BlackBerry.

The BlackBerry Connect software allows the user of the handheld device to have the BlackBerry experience of real-time push e-mail and real-time Personal Information Manager (PIM) synchronization while utilizing the handheld device's native applications.

The *BlackBerry Built-In* technology is software that is built into the ROM of a telephone that allows it to use the BlackBerry software normally running on a real BlackBerry handheld device. The end result of this is that the user can have a telephone or handheld device that is not a BlackBerry, but still enjoy the BlackBerry experience of always-on push e-mail and real-time PIM synchronization.

The advantage of these two technologies to the BlackBerry administrator is that he or she will not have to learn how to support new handheld or telephone devices. The BlackBerry Enterprise Server (BES) will treat all handheld devices the same, and all management techniques will apply, no matter what kind of hardware is being used by the end user.

BlackBerry Enterprise Server

The BlackBerry Enterprise Server (BES) is a service that works with a Lotus Domino Server, Microsoft Exchange, or Novell GroupWise mail server. The BES is a critical part of the BlackBerry infrastructure. Without it, secure access to corporate e-mail (including PIM data) would not be possible. The BES performs many functions and some of them are handled by the following BES components:

- BlackBerry Dispatcher
- GroupWise Connector (Novell GroupWise-specific)
- BlackBerry Attachment Service
- Mobile Data Service (MDS)

❏   BlackBerry Router

❏   Configuration Database

❏   BlackBerry Messaging Agent

❏   BlackBerry Synchronization Service

Let's take a look at these in more detail.

BlackBerry Dispatcher

The BlackBerry Dispatcher encrypts and compresses all data that passes between the BlackBerry hand-held units and the BES. Incoming data is unencrypted, then uncompressed, and finally passed to the appropriate BES component. Outgoing data is received from the BES component, compressed, encrypted, and passed to the BlackBerry Router to be delivered to the BlackBerry handheld unit through the BlackBerry NOC.

GroupWise Connector

The GroupWise Connector is specific to the Novell GroupWise BES. A trusted application key is generated against the primary GroupWise domain. This key is then used by the GroupWise Connector to authenticate with GroupWise. The GroupWise Connector's function is to monitor all users' mailboxes and collect PIM and e-mail changes. Any PIM or e-mail that must be sent to the handheld device is placed in a work queue table that resides in the Configuration Database. This extra table within the Configuration Database is specific to Novell GroupWise and does not exist in the Lotus Domino or Microsoft Exchange Configuration Databases.

BlackBerry Attachment Service

The Attachment Service is an NT service that takes care of handling all aspects of viewing attachments on the BlackBerry. If a user receives an e-mail that has one or more attachments, the BES does not send these attachments to the user's handheld device with the message. Instead, it places a flag in the message that indicates to the user and the operating system on the handheld device that there are attachments. In addition to the attachment flag, the BES also places the names of the attachments in the message.

If the user wants to view the attachments on the handheld unit, he or she can request to open the attachment. Upon this request, the handheld unit sends data back to the BES indicating that the user would like to open a particular attachment. The BES goes back to the user's mailbox, retrieves the e-mail, extracts the attachment into a particular directory, and asks the Attachment Service to process the attachment.

Like many other wireless devices, a BlackBerry handheld has a limited amount of memory. Therefore, the BlackBerry does not have the capacity to receive a full attachment. Because of this, it is the job of the Attachment Service to reduce the attachment down to a very small size, but still retain enough of the formatting to make it worth viewing on the handheld device.

The Attachment Service employs many techniques to achieve this. For example, if the attachment is a Microsoft Word document, the Attachment Service performs a binary conversion of the document into the *Universal Content Stream* (UCS) format. UCS is an efficient, proprietary format that is optimized for wireless delivery. It supports text, image, vector, and hybrid content. Any text content retains most of its original formatting. These rich text files are much smaller than a Microsoft Word document, since they do not contain any of the Microsoft Word formatting data. However, the resultant UCS file does contain minimal formatting, which includes bold, italic, and underline text.

After the attachment has been sufficiently processed, the Attachment Service informs the BES that it is ready. The BES takes the processed attachment and sends it to the user's handheld device with a flag that references the original message to which it was attached.

Any images embedded in the document are stripped out. However, in the case of Microsoft Word documents, a flag indicating their presence will be placed into the attachment viewed on the handheld unit. This enables the user to request these images separately if desired.

If the attachment is an image, or if the user wants to view an image that is embedded in another document, the same process will be followed. This time, the Attachment Service simply reduces the physical size of the image to conform to the BlackBerry handheld unit's screen size and resolution (160×240, 240×240, or 240×260).

Because attachments can be very big and complex, the Attachment Service can produce large CPU spikes on the server as it distills through attachments.

Mobile Data Service

Mobile Data Service (MDS) is a secure conduit that exists between all BlackBerry handheld units and their home BES. Data is sent from MDS to the BES, which sends it to the handheld units. Returning data is sent to the BES and then on to MDS.

You can also think of the MDS as an HTTP and TCP/IP proxy with special features. These special MDS features allow developers to easily push content out to the BlackBerry handheld devices. For example, a developer can write an application to send what is called a *channel* out to a handheld unit without knowing anything about the handheld unit itself. All the developer does is create two icons and a Web page, post them all on a Web server, and issue a POST command to the MDS service with certain parameters. The MDS service takes care of fetching the icons and the Web page, as well as pushing them to the handheld device as a channel through the BES.

The original name of MDS was actually *IP Proxy*, and thus the reason that the protocol used is called IP Proxy Protocol (IPPP). It is an IP Proxy, more specifically, an HTTP and TCP/IP proxy.

We will cover all the ways that you can use MDS later, because this is a very powerful component of the BlackBerry architecture. In Chapter 9, we will cover MDS in greater detail, while in Chapter 10 we will discuss how to use MDS to your advantage by building an intranet portal that is accessible from the BlackBerry. In Chapters 11 and 12, we will learn about the different special MDS features that allow you to easily push content out to BlackBerry devices.

After MDS has been enabled for a particular BlackBerry user, he or she will see a new icon on the handheld device called *BlackBerry Browser* (although this can be renamed later by the administrator). If the user uses the BlackBerry Browser to view Web pages on the BlackBerry, he or she is unknowingly using MDS. This is because the requests for the Web pages go from the handheld unit to the BES to MDS. The MDS service forwards the requests to the internal IP infrastructure to be processed.

If the request is for an internal intranet Web site, it is dealt with inside the company's intranet. However, if the request is for an external Web server, it will be forwarded out of the firewall and onto the Internet. When the data from the request returns, it goes back to the MDS, which sends it to the handheld device through the BES. This small feature alone can allow you to create an internal mobile portal that users can access from anywhere in the world.

BlackBerry Router

The BlackBerry Router is an NT service that facilitates the idea of Least Cost Routing (LCR). You can think of the BlackBerry Router as a broker. It ensures that the data is going between the handheld devices and the BES along the fastest path possible.

When the handheld device is communicating over the cellular data network, the BlackBerry Router will know this and will send data to the handheld unit over that network. If the handheld device is plugged into a USB or serial port on a PC on the internal network, the BlackBerry Router will stop routing data over the cellular network and start using the internal LAN to route data. If the handheld device uses WiFi, it will use the internal WiFi network to route all data and will facilitate all Voice over IP (VoIP) calls to and from the handheld unit.

This feature offers many advantages, especially when the BlackBerry user is communicating on a cellular network that does not implement a flat fee for data, or is communicating on a network where the user's company has chosen not to purchase a flat-fee data plan. When the handheld device is communicating on the internal network, no data is sent or received over the cellular network. This saves the user (or the company) money on cellular data charges.

This feature also allows BlackBerry administrators to provision new devices either wirelessly or on a LAN. It also allows administrators to distribute new handheld applications by either forcing the user to be connected to the LAN to receive the application, or by allowing the user a choice of whether to receive the application over the air or while connected to the internal LAN.

Configuration Database

The Configuration Database is a database that contains all configuration data for each BES component, BlackBerry user, and handheld device. Today, this database can exist on either a Microsoft SQL database server or on a Microsoft Database Engine (MSDE) server. Future releases of the BES may use IBM DB2, Oracle, or other database servers.

The information contained in the configuration database includes the data obtained during wireless backups of the handheld unit and all configurations for the user (including the handheld signature, e-mail filters, handheld ribbon positions, asset trail information about each handheld device, and more).

You can design your BlackBerry infrastructure in such a way that all of your BESs communicate with the same Configuration Database. This is advantageous because it allows you to move users between BlackBerry Enterprise Servers very easily, set a global IT policy that applies to all of your BlackBerry users, and query one database to see asset trail information on every handheld device in your organization. IT Policies are very powerful and they allow you to control every aspect of the user's handheld experience. We will discuss the IT policy in greater detail in Chapter 2.

If you design your infrastructure in this way, you must remember that currently the BES 4.0 servers do not function if you enable two-way SQL replication. This means that you must designate one SQL server to be used exclusively for all administrative functions. If the connections between your offices are too slow or too congested, you may choose to design your infrastructure so that each office has its own SQL server. If the offices are small, you may even choose to use MSDE instead of a full-blown SQL server. In this scenario, you would not be able to move users between BES servers at all. There would be no way to have a global IT Policy, but the IT Policy would have to be unique to each office. The asset trail information would pertain to only the handheld devices in that office. So, if your company needed to keep track of all handheld devices no matter which office they were in, you would need to query the Configuration Databases in each office and build one combined list manually. In future BES releases, RIM will add support for two-way SQL replication. Once this is available, the design of your BlackBerry infrastructure will be simplified.

Neither of the described implementation scenarios is right or wrong. You must determine the best fit for your organization according to your requirements and network topology.

In a Novell GroupWise environment, the Configuration Database includes a work queue. This work queue is used for all interactions between the BES and the user's mailbox. Because of this, the Configuration Database is unique to each BES and each BES can only have one configuration database. Multiple BESs cannot share one database.

BlackBerry Messaging Agent

Depending on the e-mail system, the BlackBerry Messaging Agent (BMA) either continuously scans users' mailboxes looking for new unread e-mail, scans the work queue in the Configuration Database for new items, or it receives a Messaging Application Programming Interface (MAPI) notification through a User Datagram Packet (UDP) from the mail server telling it that new mail has arrived. Once it receives the MAPI notification (or if it finds new, unread e-mail during a scan of the user's mail files or the work queue table), the BMA sends a copy of the e-mail to the BlackBerry Dispatcher. Here the first 2 KB of information is extracted, compressed down to approximately 1 KB, encrypted, and sent to the user's BlackBerry handheld unit. If that e-mail contains one or more attachments, the BMA places a special flag in the e-mail and inserts the attachment names.

If a user receives an e-mail with one or more attachments, the user can request to open the attachment. When that request is received from the user's handheld unit, the BMA goes back to the user's mailbox (or in the case of Novell GroupWise, the BMA places the request in the work queue table for the GroupWise Connector), retrieves the original message, extracts any attachments, and passes the attachments to the Attachment Service for processing. After the attachments have been processed, the BMA will send the resultant rich text of the attachments to the user's handheld unit. The attachments are formatted using the UCS format.

If the user forwards an e-mail from the handheld unit, the BMA returns to the user's mailbox, retrieves the original message (including any attachments), appends the forwarded text (if any) to the e-mail, and sends it using the normal mechanisms of the e-mail system. At the user's request, it will also place a copy of the e-mail in the user's Sent folder. In the case of Novell GroupWise, the BMA will return a request to the work queue and the GroupWise Connector will retrieve the original message.

If the user does a reply to an e-mail, the attachments are never sent. RIM decided to do this with replies because it is assumed that all of the recipients already have the attachments.

The BMA also monitors the user's calendar and looks for new entries. In the case of GroupWise, the BMA monitors the work queue looking for pending calendar entries. If the BMA finds new calendar entries, it takes a copy of the new entry and sends it to the BlackBerry Dispatcher, where it is compressed, encrypted, and sent to the user's BlackBerry handheld unit.

Since the user can create new meetings and appointments on the handheld unit, the BMA will receive this data and make the necessary changes to the user's calendar. It will also send out any meeting requests, meeting acceptances, or denials. In the case of GroupWise, the BMA places these changes in the work queue. The GroupWise Connector then makes the changes in the user's calendar.

BlackBerry Synchronization Service

The BlackBerry Synchronization Service (BSS) allows a user's Memo Pad (Journal in Lotus Notes, Notes in Outlook, or Posted Note in GroupWise), Personal Address Book, and Tasks to be wirelessly synchronized with his or her BlackBerry handheld device. Like the other BES components, the BSS sends and receives all data via the BlackBerry Dispatcher, which compresses, encrypts, and sends outgoing data, as well as de-encrypting and decompressing all incoming data.

BlackBerry Network Operations Center

All communications between the BES and the BlackBerry handheld device go through the BlackBerry NOC.

Each cellular carrier that supports the RIM BlackBerry sets up a secure Virtual Private Network (VPN) connection between the carriers' data centers and the BlackBerry NOC. All communication between the BlackBerry handheld units and the BlackBerry NOC is securely transmitted through these VPN connections.

All BlackBerry Enterprise Servers have a unique Server Relay Protocol (SRP) ID. This number identifies them to the BlackBerry NOC. When the BES sends data to the BlackBerry NOC, it is sent over the Internet as Triple Data Encryption Standard (3DES) encrypted data. 3DES (pronounced triple DES) encryption is based on the DES algorithm developed by IBM in 1974. Three 64-bit keys are used in 3DES. The data is encrypted with all three keys one after the other.

With BlackBerry 4.0, 3DES and Advanced Encryption System (AES) are now supported. AES is a block cipher algorithm adopted as a data encryption standard by the United States Government for information up to Top Secret level when it is encrypted using 192-bit or 256-bit key lengths. When you set up your BES, you can choose whether to support one or both encryption methods.

All BlackBerry handheld devices have a unique Personal Identification Number (PIN). This number identifies them to the BlackBerry NOC and the BES. All BlackBerry handheld units have a Service Book entry that identifies the SRP ID of their home BES. When a BlackBerry handheld unit is turned on, or when it comes into wireless coverage, it identifies itself to the BlackBerry NOC. At that point, the BlackBerry NOC knows how to communicate with it. More specifically, the BlackBerry NOC knows which cellular carrier the handheld unit is on, and, therefore, which VPN connection to use to communicate with the handheld device.

If a handheld device is roaming to a foreign cellular carrier, and that carrier has a GPRS Roaming Exchange (GRX) roaming agreement with the handheld device's home carrier (even if the roaming carrier does not support the BlackBerry), the handheld unit will be able to connect to the BlackBerry NOC. More specifically, the data from the handheld unit will travel to the roaming carrier's data backbone network, then to the home carrier's backbone network, and finally to the BlackBerry NOC through the VPN connection.

When the BES sends data to the user's handheld device, it already knows the PIN of the handheld unit. It addresses the data to the PIN, as opposed to the user's e-mail address or name. When the BES sends the data to the handheld unit, it actually sends it to the BlackBerry NOC. The BlackBerry NOC already knows where to find the handheld device and relays the data to it. If the handheld unit is out of coverage, the BlackBerry NOC queues the data and sends it when the handheld device is once again connected.

Data Network

The Data Network is the network that allows the BlackBerry handheld units to communicate. In the beginning, these networks were data-only networks called Mobitex (BellSouth/Cingular) and DataTAC (Motient).

As mentioned earlier, the first BlackBerrys were data-only devices. The Mobitex and DataTAC networks were (and still are) slow data networks specifically designed for small amounts of live data. In addition to BlackBerrys, they support devices such as soda vending machines that send current stock information to their manufacturers.

All data on the Mobitex and DataTAC networks are layer 2 of the Open Systems Interconnection (OSI) Model, also known as the Media Access Control (MAC) layer. This means that there is no layer 3 (also known as the Network layer) protocol such as the Internet Protocol (IP). All communication between the BlackBerry NOC and the handheld device is done using the device's MAC address, which is its unique PIN.

Later, RIM created BlackBerry devices that operated on cellular networks. The most popular and widely used cellular network by far is the GSM network. The GSM technology is used in about 70 percent of all countries around the world, and so it made sense for RIM to create GSM BlackBerry devices.

The data layer on the GSM network is GPRS. All data on the GPRS network is IP traffic, and it is this network that the BlackBerry handheld units use.

All communications on these networks is done with IP. The data networks take care of ensuring that the IP communications do not become interrupted when the handheld device is carried by the user when he or she travels. Applications on the handheld unit simply see a constant layer 3 IP network. However, as the handheld user travels between cellular communications towers, the communication data must be handed off from tower to tower while maintaining the IP connection. This is done quite elegantly in most cases. On GSM networks, which are based on Time Division Multiple Access (TDMA) technology, the handoffs are hard. On CDMA networks, the handoffs are soft. A *hard handoff* occurs when an existing connection to a base station is closed before a new connection has been opened. A *soft handoff* occurs when a new connection to a base station is opened before the previous connection has been closed. Technically speaking, it is more likely that a call or data connection will be dropped when hard handoffs are used.

Putting It All Together

Figure 1-2 shows how the different components of the BlackBerry 4.0 platform communicate with each other and the mail servers in a Lotus Domino mail environment. It also shows how the data flows out from the different BlackBerry components, through the BlackBerry Dispatcher, then to the BlackBerry Router, and finally on through the firewall and to the handheld.

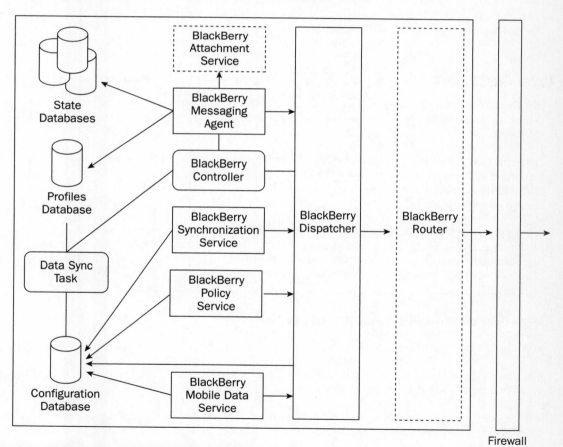

Figure 1-2: The Lotus Domino BES Components

Figure 1-3 shows how the different components work together in a Microsoft Exchange e-mail environment.

Figure 1-4 shows the different components in the Novell GroupWise e-mail environment. Note the addition of the GroupWise Connector and its interaction with the Configuration Database.

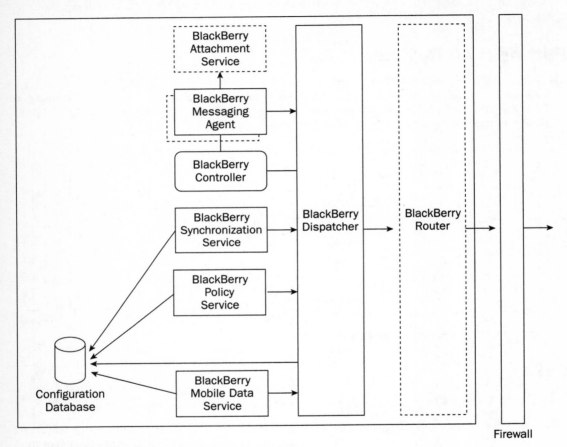

Figure 1-3: The Microsoft Exchange BES 4.0 Components

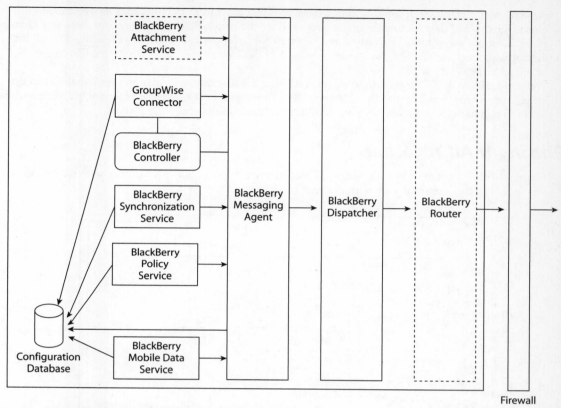

Figure 1-4: The Novell GroupWise BES 4.0 Components

Legacy Architecture (Versions 2.2 and 3.6)

When BES 4.0 shipped, it added support for Novell GroupWise and also added many new features to the BlackBerry experience. Depending on whether you were using the Lotus Domino version or the Microsoft Exchange version, the number of new features differed. However, one of the main aims of BES 4.0 was to match the feature sets between all mail platforms.

Differences Between BES 4.0 and Older BES Versions

If you are running the Lotus Domino version of the BES, you may still be using the pre-4.0 version (BES 2.2). If you are running the Microsoft Exchange version of the BES, you may be using the pre-4.0 version (BES 3.6). The Microsoft Exchange BES 3.6 has a few more features than the Lotus Domino BES 2.2.

The main differences between Microsoft Exchange BES 3.6 and Lotus Domino BES 2.2 are in the area of read/unread flags and IT Commands. While Exchange BES 3.6 wirelessly synchronized the read/unread flags for e-mails, Domino BES 2.2 did not. Exchange BES 3.6 also allowed for the use of IT Commands (which are commands that could be sent to the handheld units from the BES, such as a command to erase all handheld data, or to change the password for the handheld device if the user forgot it). Domino BES 2.2 did not allow these IT Commands. When BES 4.0 shipped, all features were offered on all BES platforms.

The pre-4.0 Domino BES infrastructure is also missing the BlackBerry Router and the capability to use an SQL server for the Configuration Database. The Configuration Database is completely absent and the Domino database (called BlackBerry Profiles) is used to store all user configuration.

Putting It All Together

Figures 1-5 and 1-6 shows how the different BlackBerry components work together with the e-mail system and how they communicate with the handheld units.

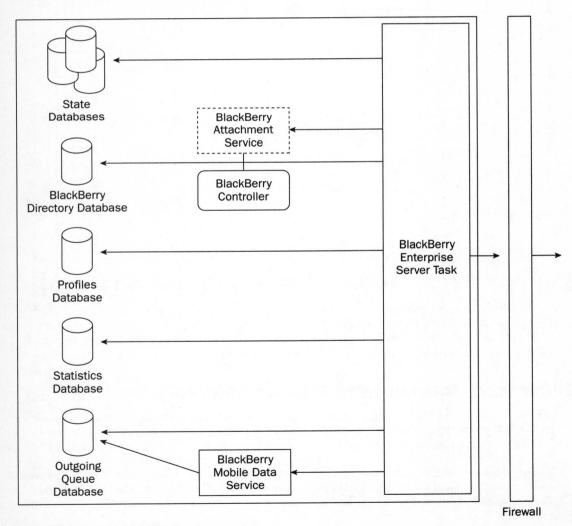

Figure 1-5: Lotus Domino BES 2.2 Components

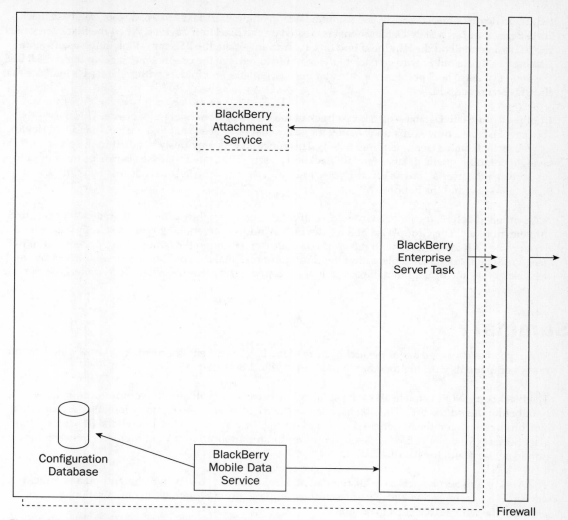

Figure 1-6: Microsoft Exchange BES 3.6 Components

Mail System-Specific Differences

No matter which e-mail system your BES is operating in, all of the components are the same. However, there are small differences in how some of the components are implemented.

In a Lotus Domino environment, the main BES component runs as a Domino add-in task. Much of the user configuration information in pre-4.0 BES is kept in a Domino database called the BlackBerry Profiles. To keep accurate track of data that is on the network (for example, in the user's mail file) and on the handheld device simultaneously, the Domino version of the BES makes use of a State Database. Each BlackBerry user has a State Database that is created on the BlackBerry server. When the BlackBerry user replies to an e-mail on the handheld unit or forwards an e-mail, the BES must look in the user's state database to figure out how to go back to the mail file to retrieve the original message. In the pre-4.0 BES, the State Database held pointers for calendar entries, e-mails, and folders within the user's mail file that the BES had processed.

In BES 4.0, the State Database also keeps track of the user's personal outgoing queue. This means that the State Database now keeps track of any messages that have been sent to the user's handheld device, as well as their status (sent, pending, failed). This function was previously handled in the pre-4.0 BES by a single outgoing queue database for all BlackBerry users. The State Database also keeps track of pointers between the user's `journal.nsf` (Notes Journal), `names.nsf` (personal address book) files, and Tasks or To-Dos and the handheld.

As mentioned before, the Novell GroupWise BES 4.0 operates slightly different than the other two BES implementations. The GroupWise BES resides on a Windows server along with the GroupWise Client 6.5.4 or later. All interactions with the user's data are done through the GroupWise Connector. The BES uses a unique work queue table within the Configuration Database (BESMgmt) for all transactions and, because of this, the BESMgmt database cannot be shared among multiple BlackBerry Enterprise Servers.

Summary

So far, you have learned about the architecture of the BlackBerry environment. You know that there are many components working together to provide the BlackBerry experience.

The BlackBerry NOC is the heart of the architecture because it facilitates the communication between the handheld unit and the BES. The BES has the most work because it is constantly looking for changes to synchronize between the handheld device and the mailbox. Working closely with the BES is the Mobile Data Service (MDS), which provides a secure conduit for applications to communicate between the corporate network and the handheld unit.

The Attachment Service acts as a "distiller" that provides a way to view attachments on a handheld device. The Configuration Database acts as the dumping ground for all user configuration data and a place for the wireless backups to be stored.

The Data Network provides the necessary transport that allows the handheld unit to communicate while the user is out of the office. Finally, the BlackBerry Router provides a "broker" service to ensure that the handheld device is always communicating with the BES using the fastest possible method.

In Chapter 2, we will cover how to plan your first BlackBerry installation. We will discuss topics such as how the BlackBerry platform communicates with your e-mail system and where to place the different BES components. We will go into detail about the IT Policy and will discuss both the new BES 4.0 and the older BES 3.6 and 2.2 versions.

2

Planning Your First BlackBerry Installation

This chapter discusses the planning and installation of your first BlackBerry implementation. We will cover how the BlackBerry platform integrates into your existing environment and some pre- and post-installation tasks. Note that you should use the installation manual to perform the actual installation.

What we cover here will help you understand the installation process a bit better and hopefully make it as smooth as possible.

Integrating BlackBerry with an Existing Infrastructure

Generally speaking, the BlackBerry solution is not very intrusive: It quietly integrates into your Lotus Domino, Microsoft Exchange, and Novell GroupWise infrastructure with ease.

As discussed in Chapter 1, the primary function of the BlackBerry Enterprise Server (BES) is to push out e-mail to the BlackBerry handheld devices. As a secondary function, the BES wirelessly synchronizes the user's calendar, memo pad, tasks, and personal address book, as well as allowing the user of the handheld device to perform global address book look-ups.

The primary and secondary functions require the BES to have the correct access control to all users' e-mail files, calendars, memo pads, and tasks. In addition, the BES must have read access to your company's global address book. This access can be provided by the use of groups (such as security groups). With Novell GroupWise, this security clearance is achieved by using a trusted application key. This key is generated so that the GroupWise Connector can access any mailbox.

A tertiary function of the BlackBerry infrastructure is the Mobile Data Service (MDS). This server itself does not require any access to your network resources. All it must be able to do is to open Web pages. If a corporate application is written that utilizes MDS to put data into a corporate database, the access control lists may need to be modified to allow this.

The Attachment Service does not need access to your network resources because its function is self-contained.

Mail Server/Post Office

The mail server (or Post Office in GroupWise) is the place where the most interaction between the BES and the e-mail system will take place. In common Lotus Domino configurations, the BES will poll each user's mailbox every 20 to 60 seconds looking for changes. Even if you increase this interval, it is common for the BES to have constant communications with the mail servers. The BES GroupWise Connector does the polling on the GroupWise BES and it polls the mailboxes every 20 seconds by default.

In Microsoft Exchange configurations, the Exchange server actually sends a UDP packet to the BES any time new e-mail has arrived for a particular user. In the Exchange environment, it is easy to see that there will be less interaction between the BES and the mail server, especially in times of low e-mail traffic. In the Domino environment, no matter how infrequent the e-mail is, the BES still polls on a regular schedule.

One of the most important aspects to remember is the way attachments are handled when BlackBerry devices are used. If a BlackBerry user receives an e-mail, the BES always makes a temporary copy of it from the user's mailbox to itself. If that e-mail has one or more attachments, then references to those attachments will be used to insert flags within the e-mail to indicate that attachments are available. If the user replies or forwards that e-mail, the BES will go back to the user's mailbox, retrieve the e-mail again (with the attachments if it is a forward), and perform the reply or forward, which will resend that e-mail with the attachments.

This kind of traffic between the BES and the mail server lends itself to a certain design methodology. For the best performance of the BES and mail server, the two servers must be on the same physical network and on the same IP subnet. This ensures that the communication is as fast as possible and is not impeded by layer 3 routing. (See Chapter 1 for a discussion about how the BES routes e-mails.)

As you can imagine, it would be a very bad idea to have a BES and a mail server communicating across a Wide Area Network (WAN) connection. This extra latency would cause the BES and mail server to slow down, because sockets and threads are kept alive much longer than expected. In addition, bandwidth would be consumed by transporting large e-mail attachments, which would be moved between the BES and mail server up to three times depending on what the user chooses to do with the e-mail.

Firewall

The corporate firewall is a critical part of any network today. The BlackBerry infrastructure works very well with firewalls and does not cause extra security compromises.

By default, corporate firewalls restrict all incoming Internet traffic by blocking all Transmission Control Protocol (TCP) and UDP ports while allowing outgoing traffic. Firewalls are also *stateful*, which means they are aware of communications sessions. This also means that if a host on your network starts communicating with a host on the Internet, the firewall recognizes this as a valid session and will allow two-way communications between the internal and external host, no matter what ports are used. The firewall will also allow new ports to be used, as long as it is all part of the same session.

The BlackBerry infrastructure always initiates the communications to the BlackBerry NOC via the BlackBerry Router using TCP port 3101. In pre-4.0 BlackBerry infrastructures, the BES did the communication directly.

In this design, the firewall configuration does not have to be changed and security compromises are not made.

Sometimes, administrators will also block all outgoing TCP and UDP ports and allow communication only on certain ports based on the applications being used to communicate with the Internet.

In these environments, the firewall must be configured to allow outgoing communications on TCP port 3101 only. Because the firewall will track communications in a stateful manner, there is no need to allow incoming TCP port 3101 traffic. The firewall administrators can also allow communications only between the BlackBerry NOC DNS name or IP address and the internal IP address of the BES server (or servers), on TCP port 3101.

Demilitarized Zone

The term *demilitarized zone* (DMZ) is from the military and is used to define a buffer zone between two enemies. In the world of computer networking, the DMZ defines a small network that sits between the Internet and the corporate network. It provides an extra buffer of protection between the corporate network and the Internet.

Typically, corporations place certain servers in the DMZ that must communicate with the Internet and the internal network. These servers would include Web servers and Simple Mail Transfer Protocol (SMTP) mail servers.

Your corporate guidelines may require that you place the BES in the DMZ. There are differing opinions on whether this move actually makes the network safer. In pre-4.0 BlackBerry environments, the BES also contains the MDS. Together, they both must have firewall ports open to communicate with the corporate mail servers, internal Web servers, and possibly internal database servers.

When using BlackBerry environments of version 4.0 or later, you can place the BlackBerry Router in the DMZ, while leaving the BES and MDS inside the corporate network.

If you place the BlackBerry Router in the DMZ, however, you may find that you will need to open up an additional TCP port (4101) between the BlackBerry Router and every PC on your network. This is because port 4101 is used to communicate with the BlackBerry Handheld Manager on the PC. (The Handheld Manager is part of the BlackBerry Desktop Software package, which is discussed in detail in Chapter 5.) This communication occurs when the handheld unit is plugged into a PC. When the handheld device is communicating on a cellular network, it communicates with the BlackBerry Router through the BlackBerry NOC. Having the ability to place the BlackBerry Router in the DMZ alone does reduce the number of ports that must be opened for the BlackBerry solution to work normally, because you don't need to also place the BES and MDS in the DMZ.

The ideal scenario is not to place any BlackBerry components in the DMZ. This makes it much easier to configure your BlackBerry environment and is a very safe scenario. However, if your corporate security guidelines require placing components in the DMZ, you can feel much safer knowing that you need to open only two ports to your internal network.

Pre-Installation Procedures

Before you install your first BlackBerry server, you must ensure that certain conditions have been met. Knowing in advance how you are going to set up your BlackBerry environment can save you a significant amount of reconfiguration and frustration later on.

Hardware and Software

First, ensure that the server hardware platform has sufficient resources to act as a BlackBerry server. The hardware and software requirements will be different depending on the underlying mail system.

Typical software requirements include the operating system versions, software version of the mail servers, and service pack levels of the server that will be hosting the BES. The following table shows the minimum software requirements for the BlackBerry 4.0 environment as of the writing of this book.

Environment	OS on BES Server	Version on BES Server	Mail Server/ Post Office	Comments
Lotus Domino	Microsoft Windows 2000 (Standard or Advanced Server) or Microsoft Windows 2003.	Lotus Domino Server 6.5.1 or later if you will be implementing wireless PIM synchronization via Roaming Users; otherwise, Lotus Domino 6.5, 6.0.1 CF2, 5.0.12, or 5.0.3.		
Microsoft Exchange	Microsoft Windows 2000 (Standard or Advanced Server) or Microsoft Windows 2003.	Exchange 5.5, 2000, or 2003 depending on your Exchange version.	Exchange 5.5 Service Pack (SP) 4, 2000 SP2, or 2003.	Any Exchange SPs that are installed on your Exchange or servers must be installed on your BES to ensure that the Exchange System Manager is kept up to date. Use the correct CDO.DLL hot fixes to obtain the correct MAPI version.

Environment	OS on BES Server	Version on BES Server	Mail Server/ Post Office	Comments
Novell GroupWise	Microsoft Windows 2000 (Standard or Advanced Server) or Microsoft Windows 2003	Novell GroupWise client 6.5 SP4	Any GroupWise Post Office that contains any BlackBerry users must be running version 6.5 SP1 or later	

Typical hardware considerations include the number and speed of the CPUs in the server, and amount of memory installed. As of the writing of this book, the minimum and recommended hardware requirements for the BlackBerry 4.0 environment are shown in the following table.

Environment	Number of Users on the BES	Minimum Required Hardware and Configuration	Recommended Hardware and Configuration
Lotus Domino	500	Intel Pentium IV processor (2 GHz or higher); 1.5 GB of RAM; MSDE on the same computer	Intel Pentium IV processor (2 GHz or higher); 1.5 GB of RAM; MSDE on the same computer
	1,000	Dual Intel Pentium IV processors (2 GHz or higher); 3 GB of RAM; MSDE on the same computer	Dual Intel Pentium IV processors (2 GHz or higher); 2 GB of RAM; SQL on a separate server
	2,000	Dual Intel Pentium IV processors (2.8 GHz or higher); 3 GB of RAM; MSDE on the same computer	Dual Intel Pentium IV processors (2.8 GHz or higher); 2 GB of RAM; SQL on a separate server
Microsoft Exchange	500	Intel Pentium IV processor (2 GHz or higher); 1.5 GB of RAM; MSDE on the same computer	Intel Pentium IV processor (2 GHz or higher); 1.5 GB of RAM; MSDE on the same computer
	1,000	Dual Intel Pentium IV processors (2 GHz or higher); 2 GB of RAM; MSDE on the same computer	Dual Intel Pentium IV processors (2 GHz or higher); 1 GB of RAM; SQL on a separate server
	2,000	Dual Intel Pentium IV processors (2.8 GHz or higher); 4 GB of RAM; MSDE on the same computer	Dual Intel Pentium IV processors (2.8 GHz or higher); 2 GB of RAM; SQL on a separate server

Table continued on following page

Environment	Number of Users on the BES	Minimum Required Hardware and Configuration	Recommended Hardware and Configuration
Novell GroupWise	500	Intel Pentium IV processor (2 GHz or higher); 1.5 GB of RAM; MSDE on the same computer	Intel Pentium IV processor (2 GHz or higher); 1.5 GB of RAM; SQL on a separate server

Security

In pre-4.0 BES environments, all components resided on the same server as the BES itself, with the exception of the Attachment Service. However, with BES 4.0, you can split up the BlackBerry components onto different servers. With this in mind, you must plan where you will install your BlackBerry Router.

You may need to work with your corporate security team to iron out special change-management procedures, or if corporate guidelines require you to place the BlackBerry Router in the DMZ. Whatever decision is reached, you must ensure that the BlackBerry Router can make an outbound connection to the BlackBerry NOC on TCP port 3101. As discussed earlier, if the BlackBerry Router must reside in the DMZ, you will need to ensure that it can communicate with all PCs on your internal network on TCP port 4101.

Mail System Specifics

In a Lotus Domino environment, you must install the Domino software on your BlackBerry server and ensure that it always has a current replica of the Domino Directory (the names.nsf file). If you will be performing the Domino installation, you must request a new server ID file from your Domino administrators. The server ID file is needed during the installation of the Domino server and is necessary to run correctly after that. Once the Domino server has been installed and is functioning correctly, you can proceed with the BES installation.

In a Microsoft Exchange environment, ensure that you have the latest version of the Microsoft Collaboration Data Objects (CDO) dynamic link library (DLL) file installed on all your Exchange servers, including the one onto which you plan to install the BES.

In general, it is a good idea to limit the number of mail servers to which one BES connects to. Having the BES connect to too many mail servers will degrade the performance of the mail servers and the BES server because of the latency that is introduced.

When installing a Novell GroupWise BES, you first must create a GroupWise account that the GroupWise Connector will use to communicate with the mailboxes. During the installation process, you will use a Trusted Application Generator that is supplied with the GroupWise BES. This tool will allow you to generate a Trusted Application Key by pointing to the Primary GroupWise Domain.

In addition, you are required to install the US-English GroupWise client version 6.5 4 (SP4) on the server where the BES will be installed. After the installation, the GroupWise Connector will use the GroupWise account that you created in conjunction with the Trusted Application Key to communicate with the mailboxes.

Using MSDE or SQL Databases

Depending on which e-mail system you were using, the BlackBerry server in pre-4.0 environments used either a local Microsoft Database Engine (MSDE) or Lotus Domino database to store configuration information for each BlackBerry user. In the BES 4.0 environments, you have the choice of either MSDE or SQL Server to store the configurations.

If you will be conducting a pilot of BlackBerry and you plan to completely re-install everything afterward, you may choose to use MSDE. If this is going to the first of many BlackBerry servers, you may want to consider using SQL Server.

There are many advantages to using SQL Server in a multi-BES BlackBerry environment, but there are some limitations and considerations that you must be aware of before you embark on your implementation.

If you have (or plan to have) many BlackBerry servers in your office (or in multiple offices around the world), configuring them all to use one instance of an SQL database is advantageous. Remember that if you are in a GroupWise environment, the GroupWise BES cannot share the Configuration Database (called the BESMgmt database). This is because it has an extra table that is a work queue. Because the work queue is specific to the BES, each BES must have its own SQL or MSDE database.

Using this configuration means that all asset trail information is stored in a central database. This enables you to have access to information about every handheld device in your organization. It also means that you can very easily move BlackBerry users from one BES to another without the need to inconvenience the user. Another advantage becomes even more apparent when you consider users who move from one office to another for months (or years) at a time. It is also useful in the event of a disaster where you can simply move users from a BES that is down to a BES that is up in a disaster-recovery area. (Chapter 8 provides more information on disaster recovery.)

Finally, this configuration enables you to create IT Policies that can be applied on a global scale. (See the section "IT Policies" in Chapter 3 for more information about IT Policies.) Because you are using a single instance of the database, each IT Policy that you create is immediately available to be assigned to all users, no matter where they are and which BES they are on. If a change must be made to an IT Policy, it can be made in one place instead of many, simplifying administration.

Before you decide to make use of a single database instance, you should bear in mind that, while SQL supports two-way replication, the BlackBerry Enterprise Server cannot function when two-way replication is used. This means that you can set up one-way SQL replication only, and this will affect the way in which you plan your BlackBerry environment. It is still possible to make use of a single SQL database instance, but this will depend heavily on your network environment.

One way to work around the one-way replication issues would be to make use of only one SQL Server in a central location and have all BlackBerry servers pointing to it, even if communication is over a WAN connection. Available bandwidth between offices would become more critical in this scenario because each BES must be able to contact the database at will, and interruptions to this communication could cause problems.

Another approach is to set up "regional" SQL databases. For example, you could have an SQL database for all BES servers in the Americas, or even in Europe. Depending on available bandwidth and the number of BES servers used, you may need to reduce this to smaller regions (for example, U.S. East Coast, U.S. Central, U.S. West Coast, and so on). You may go even smaller, breaking the regions into states or provinces.

Your strategy, then, would be to have all BES servers in that particular region connected to the one SQL database. This would allow you to seamlessly move BlackBerry users between those BES servers. It would also allow you to do one-way SQL replication to an SQL server in a disaster-recovery site for use in the event of a disaster.

If your available bandwidth between offices is too low to support either of the scenarios described here, you can choose to install SQL Server in each office and use a different Configuration Database for each office. In smaller offices, you may want to use MSDE.

Summary

The BlackBerry architecture fits into your current environment any way you want it to and doesn't cause unnecessary security risks. The BES (or the BlackBerry Router) initiates outgoing communication to the BlackBerry NOC, which means that you don't necessarily need to open ports on your firewall.

You can choose to place components of the BlackBerry environment in a DMZ if your company requires it, although it is quite safe to locate them all solely on your corporate network.

It is important to keep your BES as close as possible to your mail server or servers (ideally on the same IP subnet for optimum performance).

There are also a few important factors that need to be worked out before you install your first BlackBerry Enterprise Server. You must ensure that your hardware platform has the sufficient resources to handle a BES. You also must ensure that you are using the correct operating system version (and patch level), and that your mail servers are operating with the correct software version.

You must plan for the placement of your BlackBerry Router or BES with your security team to ensure that the correct TCP ports are open to allow for seamless communication with the BlackBerry NOC and the PCs on your network.

Finally, you have to make the important decision on how you are going to implement the Configuration Database and how your BlackBerry Enterprise Servers and administrators will connect to it.

In Chapter 3, we are going to discuss how to deploy the BlackBerry Desktop Software if you need it in your environment. We will cover many ways to automate this process, including creating silent installs.

3

Deploying the Desktop Software

In a full BlackBerry 4.0 environment, there is no reason to install any BlackBerry software on the users' desktops because everything happens over the wireless network. However, you may need to install the Desktop Software if you synchronize the handheld devices with an application other than Microsoft Outlook, Lotus Notes, or Novell GroupWise. That said, upgrading the handheld operating system does require a physical connection to a PC, and we will discuss how this can be achieved later in Chapter 5.

If you are not yet running in a full BlackBerry 4.0 environment or are running in a pre-4.0 environment, then you will need to deploy the Desktop Software to the users' desktops. Finally, even if you are in a full BlackBerry 4.0 environment, you may want to deploy the Handheld Manager to each PC so that you can save on your cellular bills and allow the users to charge their BlackBerry devices while they are connected to their desktops.

This chapter examines the deployment of the BlackBerry Desktop software and provides tips on an automated installation. We also examine the all-important concept of IT Policies.

Desktop Software Components

When RIM first created the Desktop Software, it was called the Desktop Manager. In those days, the Desktop Manager was an all-in-one solution. As the needs of the wireless carriers and end users grew, RIM decided that the software for the handheld devices needed to be a separate component.

When BlackBerry 4.0 was released, the Desktop Software was split up even further. The Desktop Manager is now no longer the main program on the desktop, so the bundle of software that you install is now called the BlackBerry Desktop Software. Inside this bundle, you will find the Desktop

Manager along with the Application Loader and Handheld Manager. The software is broken out in this way because you now do not need to install the Desktop Manager unless you are synchronizing with applications other than Lotus Notes, Microsoft Outlook, or Novell GroupWise.

You can choose to install the Handheld Manager only, and nothing else, which will allow users to charge their handhelds and have the data sent over the LAN when the handheld is attached to the desktop. If you want users to load their own Handheld Software and third-party applications, you can install the Application Loader. Many different combinations can be employed by you, and this is possible because the software is now so flexible.

Desktop Manager

The main function of the Desktop Manager is to synchronize the data between the handheld unit and the application software on the user's desktop (including Lotus Notes, Microsoft Outlook, and Novell GroupWise). The Desktop Manager uses PumaTech's Intellisynch software to provide this functionality. The Desktop Manager communicates with the handheld device by using the Handheld Manager (see the section "Handheld Manager" later in this chapter).

In addition to this function, the Desktop Manager allows the user to make local backups of his or her handheld device, which can later be restored if data is accidentally deleted from the handheld unit. The Desktop Manager also provides an icon that launches the Application Loader (see the following section, "Application Loader"). Redirector Settings allows users to make changes to the BlackBerry (such as changing the signature, the mail filters, and so on).

Finally, the Desktop Manager informs the BES when the handheld unit is attached to the desktop. Depending on how the user's BlackBerry account has been set up, the BES may stop sending e-mails to the handheld device.

Application Loader

The Application Loader allows the user to install new firmware, a handheld device operating system, and third-party BlackBerry applications onto his or her handheld unit. The Application Loader does need the actual Handheld Software (see the section "Handheld Software" later in this chapter) installed, as well as third-party applications installed on the desktop before it is executed.

When the Application Loader is executed, it finds any Handheld Software and third-party BlackBerry applications installed on the PC. The Application Loader allows the user to choose what to install and offers to make a backup of the handheld device before it starts.

Handheld Manager

The Handheld Manager is a small piece of communications software that acts as a proxy between the handheld device and the BlackBerry Router. When the handheld device is connected to the desktop, the Handheld Manager informs the BlackBerry Router that the handheld unit with a particular PIN has connected to a desktop. At that point, the BlackBerry Router starts routing all data to and from the handheld device through the desktop. In addition to this, the Handheld Manager uses its Universal Serial Bus (USB) driver to charge the handheld unit.

The Handheld Manager is not tied to a particular user or e-mail client. It operates totally on its own. Because of this, many handheld devices can be connected to one desktop. Thus, for example, if you install a PC with a USB hub in a conference room, your users who are traveling from other offices can connect their BlackBerry devices, have them charged, and receive updates over the LAN instead of over the wireless network.

Handheld Software

The Handheld Software is a bundle that you receive from your wireless carrier. Each wireless carrier offers a different piece of Handheld Software per handheld device. The Handheld Software contains the firmware, operating system for the handheld device, and standard applications.

When Desktop Manager 3.6 was released, it no longer contained any Handheld Software. Each wireless carrier provided the Handheld Software, and it was up to the end user to download the appropriate Handheld Software from the Web site of the wireless carriers. The Desktop Manager needed to be installed first, and then the Handheld Software for the various handheld devices that would be supported.

While many wireless carriers actually sell the same models of BlackBerry, they like to customize the Handheld Software by adding their company logo to the standby screen, or other such customizations. To make use of the Handheld Software, you must install it onto the user's desktop and load it onto the handheld device using the Application Loader.

Automating the Desktop Software Installation

There are a few ways to automate the deployment of the BlackBerry Desktop Software, including the following:

❑ At each desktop, performing the steps manually.

❑ Creating a silent install that uses an answer file.

❑ Using the Microsoft Installer (MSI) file to create a Novell ZENworks object or Microsoft Systems Management Server (SMS) object for distribution.

Let's take a closer look at the last two items in that list.

Silent Install

The silent install method is required if you want to automate the BlackBerry Desktop Software installation in BlackBerry Desktop Software environments prior to version 4.0, and optional when installing BlackBerry 4.0 Desktop Software.

The BlackBerry desktop silent install entails using a two-step approach:

1. Install the Desktop Software using a special command-line switch to create an answer file.

2. Run the Desktop Manager with another special command-line switch that saves your configuration settings into three Extensible Markup Language (XML) files.

Before you decide to use the silent install method, you must ensure that your regular users have the correct local permissions in Windows. They will need the ability to write files to the Windows hard disk, have the ability to register Component Object Model (COM) components, and the ability to modify the registry, specifically the following keys:

```
HKEY_LOCAL_MACHINE\Software\Microsoft\Windows\Current Version\App Paths
HKEY_LOCAL_MACHINE\Software\Research In Motion\BlackBerry
HKEY_LOCAL_MACHINE\Software\Microsoft\Windows\Current Version\Uninstall
```

Before you begin, you must get the install file into its separate components.

If you have downloaded the Desktop Software from a Web site, it is likely compressed into an executable ZIP file or some other compressed file. To decompress it into its separate components, simply copy the file to a specific directory on your hard drive (we will use `c:\temp\blackberry`) and execute it. The file will be decompressed automatically and the setup program will start. When it asks you if you want to install, click the Cancel button, as shown in Figure 3-1. You should find the decompressed contents of the original file in the directory.

Figure 3-1: Canceling the option to install

If you have the original install files on a CD-ROM from RIM, you will not need to decompress them (they are already decompressed on the CD-ROM). You will simply need to copy them to a directory on your hard disk (in our example, to `c:\temp\blackberry`).

Creating the Setup Configuration File

The first step is to create or modify the setup configuration. Depending on which version of the BlackBerry Desktop Software you are using the methods will be different. If you are using a Desktop Manager prior to version 4.0, then you must create a `setup.iss` file. If you are using Desktop Software 4.0, you must edit the existing `setup.ini` file and add in some optional parameters.

Pre-4.0 Desktop Manager

When the files are decompressed on your hard drive, run the setup program from the command line as follows:

```
setup.exe /r /f1C:\Temp\BlackBerry\setup.iss
```

This command uses the /r switch to place the setup program into record mode and the /f1 switch to specify where to write the answer file. It is important to note that there is no space between the f1 switch and the path to the answer file.

You must not click the Back button, only the Next button. Clicking the Back button will create problems when you use the answer file to "play back" the install. As with any command-line text, if there are any spaces in the file names or path names, you must encapsulate the full path in quotation marks.

As the setup program runs, normally you must make all of the necessary choices to customize the users' desktops, as shown in Figure 3-2.

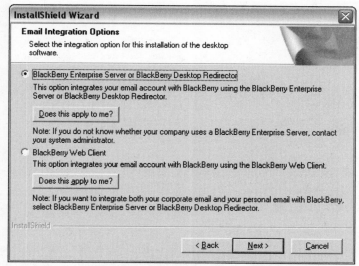

Figure 3-2: Example of the choices you will make during the install to customize the users' desktops

Once you have completed the install, the configuration file should be in the c:\temp\blackberry directory. If you open this file using the Notepad application, you should see something like the following:

```
[{01E96A75-5234-46AC-AA08-1723A00C902F}-DlgOrder]
Dlg0={01E96A75-5234-46AC-AA08-1723A00C902F}-SdWelcome-0
Count=13
Dlg1={01E96A75-5234-46AC-AA08-1723A00C902F}-SdCountrySelection
Dlg2={01E96A75-5234-46AC-AA08-1723A00C902F}-SdLicense-0
Dlg3={01E96A75-5234-46AC-AA08-1723A00C902F}-SdRegisterUser-0
Dlg4={01E96A75-5234-46AC-AA08-1723A00C902F}-SdSetupTypeSelect
Dlg5={01E96A75-5234-46AC-AA08-1723A00C902F}-SdAskOptions-0
```

```
Dlg6={01E96A75-5234-46AC-AA08-1723A00C902F}-SdMailServerSelection
Dlg7={01E96A75-5234-46AC-AA08-1723A00C902F}-SdAskOptions-1
Dlg8={01E96A75-5234-46AC-AA08-1723A00C902F}-SdRedirectorSelection
  ...
```

Desktop Software 4.0 or Later

The Desktop Software 4.0 comes with a file that controls the way in which the software is installed. You can edit this file and make changes to it, including adding in some optional parameters. Once the installer has been decompressed into its separate components, you will see a file called `setup.ini`. If you open this file using the Notepad application, you will be able to modify it, depending on how you want the install to proceed. Following is a snippet of what you might see in the `setup.ini` file:

```
[Info]
Name=INTL
Version=1.00.000
DiskSpace=8000   ;DiskSpace requirement in KB

[Startup]
CmdLine=/l*v %TEMP%\BB_DM.log
SuppressWrongOS=N
ScriptDriven=2
ScriptVer=9.0.0.333
  .
  .
  .
[INSTALL_OPTIONS]
COUNTRY =  New Zealand
SETUPTYPE = Enterprise
MAILSERVER = GroupWise
```

You can add a new line at the end of the file tag called [INSTALL_OPTIONS]. Under that line you can add the new configuration parameters shown in the following table.

Installation Option	Action	Default
COUNTRY	Specify the countries to display the software license agreement for.	United States
SETUPTYPE	Specify if the setup type is Enterprise or Internet.	Enterprise
MAILSERVER	Specify if Exchange, Lotus, or GroupWise is the messaging and collaboration server used by the user.	Exchange
REDIRECTOR	Specify if the Desktop or Enterprise controls e-mail redirection.	Enterprise

Installation Option	Action	Default
USERS	Specify if AllUsers or CurrentUser has access to the Desktop Software.	AllUsers
SHORTCUTSTARTUP	Specify whether to add a Desktop Software shortcut to the Start menu.	False
SHORTCUTDESKTOP	Specify whether to add a Desktop Software shortcut to the desktop.	False
HANDHELDINSTALLERDIR	Specify the path to the Handheld Software installation files if you do not want to maintain Handheld Software in a shared software directory.	Device folder in the network directory (for example, *<drive:\network directory>* Device, where *drive* indicates your drive location and *network directory* indicates the directory on the network)
UNINSTALLEXISTINGHANDHELDS	Specify whether to remove existing Handheld Software bundles during the installation.	False
SUPRESSREBOOT	Specify whether to suppress any reboots that might occur during the installation.	False

Be careful not to create conflicting configurations. If you do have conflicting configurations, the configuration option that came first will be used and the conflicting option that comes second will be ignored. For example, if you set the SETUPTYPE to Internet, and the REDIRECTOR to Enterprise, the REDIRECTOR option will be ignored.

You can now run the installer and allow it to install the Desktop Software onto your computer.

Creating the Default Configuration XML Files

Now that you have the Desktop Software installed, you must run the Desktop Manager with a special switch that tells it to record all your selections into three XML files. From the command line, run the Desktop Manager as follows:

```
DesktopMgr.exe /r C:\Temp\BlackBerry
```

The /r switch tells the Desktop Manager to record your settings into the XML files and the path tells it where to save the XML files.

Use the Desktop Manager to navigate through its menus and to make all the changes you want to be the defaults on all your users' desktops. When you are finished, close the Desktop Manager. You will see the following three XML files in c:\temp\blackberry:

❑ defaultsdesktop.xml

❑ defaultsintellisync.xml

❑ defaultsredirectorsettings.xml

Following is an example of the contents of the defaultsdesktop.xml file:

```xml
<?xml version="1.0" encoding="UTF-8"?>
<Default_Desktop_Settings>
  <key name="Software">
    <key name="Research In Motion">
      <key name="BlackBerry">
        <key name="BackupRestore">
          <value name="Number Of Days" type="REG_DWORD">7</value>
          <value name="Auto Backup Enabled" type="REG_DWORD">1</value>
          <value name="Auto Backup Exclusions" type="REG_DWORD">0</value>
        </key>
        <key name="Manager">
          <key name="Settings">
            <value name="DialogSize" type="REG_DWORD">25428196</value>
            <value name="HideWhenMinimized" type="REG_DWORD">1</value>
            <value name="RememberMyPassword" type="REG_DWORD">0</value>
            <value name="ConnectionSetting" type="REG_DWORD">4294967295</value>
            <value name="MaximumSerialPortSpeed" type="REG_DWORD">4</value>
            <value name="DisableDevicePowerSave" type="REG_DWORD">0</value>
            <value name="DisableScreenSaverDetection" type="REG_DWORD">0</value>
            <value name="PasswordHandheldPrompt" type="REG_DWORD">0</value>
            <value name="ExtensionViewMode" type="REG_DWORD">0</value>
            <value name="ShowStatusBar" type="REG_DWORD">1</value>
            <value name="ShowPolicyErrMsg" type="REG_DWORD">1</value>
            <value name="modemdevicepin" type="REG_DWORD">0</value>
          </key>
        </key>
        <key name="Synchronize">
          <value name="ViewLog" type="REG_SZ">Notepad.exe</value>
          <value name="Parameters" type="REG_DWORD">58</value>
        </key>
      </key>
    </key>
  </key>
</Default_Desktop_Settings>
```

Using the Silent Install

After you have the three XML files and the setup configuration file in the proper directory, along with the rest of your install files, you are ready to use them. If you are in a pre-4.0 Lotus Domino or pre-3.6 Microsoft Exchange BlackBerry environment, and you have created an IT Policy that you would like to distribute with the install, copy the policy.bin file into the directory along with the rest of the files. (See the section "IT Policies" later in this chapter for more information on creating IT Policies.) If you would also like to distribute the Handheld Software files along with the install, simply create a new directory called device (for example, c:\temp\blackberry\device) and copy the executables for each Handheld Software into that directory. These handheld-specific software files can be downloaded from your carrier's Web site. Do not install them, but simply place the exe files into the device directory. When the install process is eventually run on the user's PC, the installer will see the device directory and execute each of the handheld software files to install them.

To make these files available for everyone, you should copy them to a location on one of your network volumes. Be aware that, under certain circumstances when using a Novell NetWare network, long file names and path names are truncated, which may disrupt the installation process if the user is running the install from a network volume. If this happens, compress all of the files to a ZIP file while retaining the path names. Then, uncompress that ZIP file to the user's hard drive and run the setup program from there.

Now that you have the files available for everyone, you can choose your method of distribution. You could launch the installation using a login script. A second option is to send everyone an e-mail with a link to the Unified Naming Convention (UNC) share and ask them to run a file. A third option is to create a Novell ZENworks object, which simply launches the file for the user. This is a choice that you will need to make based on the comfort level you have with your users and their expectation levels.

However you choose to have this installed on the users' computers, you must know the command line that will be used when running the setup program to make it use the configuration file.

Pre-4.0 Desktop Manager

The command line that you must use is as follows:

```
C:\Temp\BlackBerry\setup.exe /s /f1C:\Temp\BlackBerry\setup.iss
```

Here you are running the setup program and using the /s switch to tell the program to use an answer file. The /f1 switch tells the setup program where to find the configuration file. Note that there is no space between the switch f1 and the path to the answer file.

When this command is executed, the setup program will perform the install silently. This means that the users will not see any dialog boxes on their screens during the install.

If you want to track how the install has performed on each PC, you can use the /f2 switch to cause the setup program to save a log file. The command line would then look like this:

```
C:\Temp\BlackBerry\setup.exe /s /f1C:\Temp\BlackBerry\setup.iss
/f2"%TEMP%\bbinstall.log"
```

Here the /f2 switch is followed by a path to the log file. Note that there is no space between the /f2 switch and the path to the log file.

Of course, the examples shown here presume that you are running the install from c:\temp\blackberry, but you may have chosen to copy the files to a network volume. In that case, you would simply replace c:\temp\blackberry with the appropriate path to your install files.

If you have different groups of users who require different default settings, you will need to repeat this process and save the files in a different location on your network. You can then use the different version as you want.

Normally, the silent install technique is used for fresh installations, but you can use it for performing upgrades to older versions of the BlackBerry Desktop Software.

Desktop Software 4.0

When you are using the Desktop Software version 4.0, you have a few extra choices on how you want the silent install to proceed. You can specify the language to be used by the installer (for example, if it is not English) and the components to install. The following table shows the options for the Language ID and Installation Level ID. The Installation Level ID is simply a number that corresponds with a combination of Desktop Software components.

Language	Language ID
French	1036
German	1031
Italian	1040
Spanish	1034

Desktop Software Configuration	Installation Level ID
Handheld Manager only	1
Handheld Manager and Desktop Manager	100 (Default)
Handheld Manager, Desktop Manager, and Certificate Synchronization	125

So, to implement a silent install, you must run `setup.exe` using the `/s` parameter, plus one (or both) of the optional parameters for language and installation components. The setup command will be in the following format:

```
<Drive:\Network Directory>\setup.exe /s /l<Language ID>/v"INSTALLLEVEL=
<Install Level ID>"
```

Following is an example:

```
C:\Temp\BlackBerry\setup.exe /s /l1040 /v"INSTALLLEVEL=1"
```

This command assumes that you are still running the setup from `c:\temp\blackberry`, but copying all of the files to a network volume would make it available to all of your users. The example command above specifies the `/l` parameter with a value of `1040` (which is `Italian`). If you omit the `/l` parameter, English will be used. The example also uses the `/v` parameter with the value of `INSTALLLEVEL=1` (which instructs the installer to install only the Handheld Manager). If you omit the `/v` parameter, the installer will install the Handheld Manager and the Desktop Manager.

Using MSI Files

Microsoft Installer (MSI) files have since been renamed to Windows Installer files and are now the standard for installing applications onto Windows computers. We will not go into detail about MSI files in this chapter because the subject itself can be discussed in great depth in a book of its own. We will assume that you have a working knowledge of MSI files and that you have worked with them before.

In pre-4.0 versions of the BlackBerry Desktop Software, RIM never supplied an MSI file. If you wanted to use the MSI installation method, you had to create your own MSI file. You can create your own MSI files by using a program that takes a snapshot of your computer before and after you install the Desktop Manager. An example of a program that has this kind of functionality is `WinInstall LE 2003` (which you can download from `www.ondemandsoftware.com/SP_Download/wiLE2003.asp`). This is a free program, but it has enough functionality to create your installer MSI.

With the release of the 4.0 version of the BlackBerry software, RIM has provided a full installer MSI file.

The installer MSI file can be used to create a distributed application with programs such Novell ZENWorks and Microsoft Systems Management Server (SMS).

IT Policies

The IT Policies are sets of parameters that a BlackBerry administrator can use to enforce company policy, security, or simply to limit the available functions on the BlackBerry handheld device.

It is important to sit down with your security, operations, marketing, and business teams to discuss the different IT Policy options available for the BlackBerry. It is likely that these different teams of people will have differing ideas on how to implement the BlackBerry IT Policy. Together, you must mold the policy into something that provides a comfort level for everyone, and one that will not inconvenience your users. It is important to strike the perfect balance between ensuring your network and its end points (in this case, the BlackBerry devices) are secure enough so that your users' data is protected, but not make it too strict that users will shy away from using the BlackBerry.

With the BlackBerry 4.0 platform, creating, editing, and deploying IT Policies is very simple. In previous versions of the BlackBerry platform, the task was more complicated. We will discuss both methods.

BES 4.0

When using BlackBerry 4.0, you work with IT Policies either by using the BlackBerry Manager's Microsoft Management Console (MMC) snap-in in an Exchange environment, or the BlackBerry Management Console in a Domino or GroupWise environment.

With this version of the BlackBerry platform, once you have created and assigned a policy to a BlackBerry user, that policy will be sent to the user's handheld device over the air (OTA) and it will take effect immediately.

To create, modify, and assign IT Policies, launch the BlackBerry Manager application and click the BlackBerry Domain on the left of the screen, as shown in Figure 3-3. On the right side of the next screen, click Edit Properties. On the following screen, click the IT Policy shown on the left, and then click the ellipsis next to the IT Policies field. You will then see a list of IT Policies, as shown in Figure 3-4. If this is a new BES 4.0 installation, you will see the Default policy. Figure 3-5 shows a list of IT Policy Options.

At this point, you can create your own IT Policies based on your projections for future use. If you already have IT Policies assigned to users, any changes you make will be wirelessly deployed immediately.

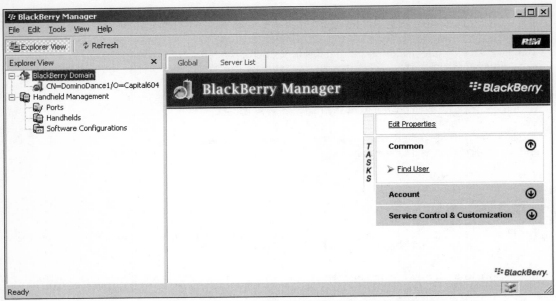

Figure 3-3: The BlackBerry Domain as seen in the BlackBerry Manager application

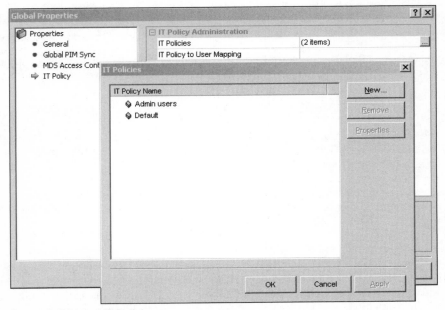

Figure 3-4: List of IT Policies

The BES 4.0 IT Policy has many more policy keys than the older BES 2.2 IT Policies. These policies are much easier to deploy because they get to the handheld units over the wireless network. In addition, any IT Policies that you create using this interface can be deployed to any BlackBerry user who is registered with the BES that shares the same SQL database.

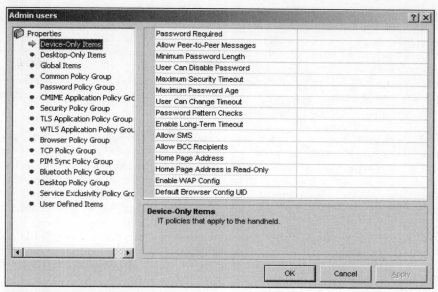

Figure 3-5: List of IT Policy Options

Some of the kinds of policy keys are Boolean (True or False), string, and integer values.

To assign an IT Policy to a user, use the BlackBerry Manager to open up the user's properties and click on IT Policy. On the right of the screen shown in Figure 3-6, choose the correct policy. To assign the same policy to multiple users, select multiple users and click the Edit Properties link in the lower half of the screen. Any changes you make will apply to all of the selected users.

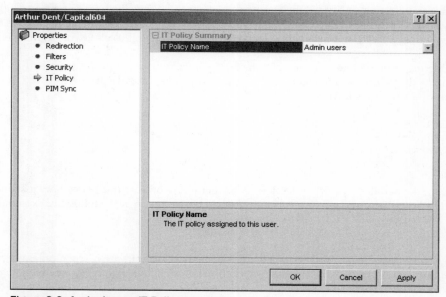

Figure 3-6: Assigning an IT Policy to a user

Domino BES 2.2 and Exchange BES 3.5

When using BlackBerry 2.2 for Domino, you must use an external tool called `itpolicy.exe`. This tool enables you to create an IT Policy file (which is a text file with the extension of `.inf`) and then compile that file into a binary file called `policy.bin`. When the `policy.bin` file is ready, you must copy it to the folder on every PC where the BlackBerry Desktop Manager resides. When the user restarts the BlackBerry Desktop Manager, the policy file will be in effect. After the BlackBerry synchronizes in the cradle, the policy will take effect on the BlackBerry.

You can create many different `.inf` files that have slightly different variations on an original IT Policy. You may want to distribute these variations to different groups of people. Unfortunately, the compiled version of the file is always called `policy.bin`, and so you must create a directory structure on a shared volume to store the different variations of the `policy.bin` file. You should keep these different versions around in the event that a computer must be rebuilt and so that the technician rebuilding the computer can copy the correct `policy.bin` file. This inflexibility often causes people to just implement one policy.

You will find the IT Policy program and its components on the BES installation CD-ROM, or in the decompressed BES installation files that you downloaded from the RIM Web site. Create a copy of the `itpolicy` directory on your computer. Inside the `itpolicy` directory, launch the program called `itpolicy.exe`.

Click the File Open button (the second icon from the left) and open the sample `policy.inf` file that you see. Figure 3-7 shows what the IT Policy program looks like with this file open. The sample `policy.inf` file shows every IT Policy setting that you can change. To figure out the parameters for each option, use the drop-down menu in the middle of the toolbar near the top of the screen to select a policy key. When the policy key is selected, the grayed out fields to the right of the drop-down menu display the type of data that the key expects and the possible values that you can assign to it.

As with BES 4.0, some of the kinds of policy keys are Boolean (True or False), string, and integer values. Some policy keys enable you to use existing Windows Registry key values. For example, the `OwnerName` key enables you to type in * to tell the Desktop Manager to retrieve the user's name out of the Windows Registry, instead of using a value in the policy file. In this case, the Desktop Manager looks for the Windows owner's name and inserts that value into the handheld device's Owner Name field.

When entering string values (for example, the `OwnerInfo` key), you must be aware of the character length of the BlackBerry screen and place the appropriate line feeds. To place a line feed into a string value, use the \ sign as shown here:

```
OwnerInfo = I'm still really upset\
about earth being destroyed.\
I think I'll have\
a nice cup of tea.
```

If you want the handheld device or Desktop Manager to use one of your policy keys just once and then ignore it in future, you add `{default}` to the key name, as shown in the following example:

```
OwnerInfo {default} = Please return to RIM
```

You would use this extra parameter in situations where you want to set a certain policy key once with a default value, and possibly allow the user to modify it, but not overwrite it every time you deploy a new `policy.bin` file.

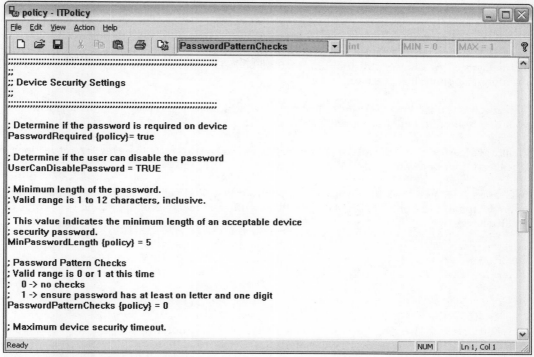

Figure 3-7: Viewing the policy.inf file

If you want to reset a certain policy key each time you deploy a new `policy.bin` file, add `{policy}` to the key name, as shown in the following example:

```
AllowDesktopAddIns {policy} = true.
```

Selecting the policy key does not add it to the file. It is merely a way to visually display how each key works so that you can manually type the key and its values into the `.inf` file.

IT Policies and Handheld Units

You should be aware of how the handheld unit deals with the IT Policies. Once a handheld device has received a wireless IT Policy OTA, it will reject the older kind of policy (that is, the kind deployed via the `policy.bin` file). If you come across a handheld device that will not accept your `policy.bin` IT Policy, it is very likely that it once belonged to someone who was registered on a unit with Domino BES 4.0 installed or a unit with Exchange BES 3.6 or higher installed, and received an OTA policy.

It is also a very good idea to make the changes to your IT Policy before you deploy it to all of your users. A good first step is to set the IT Policy to force users to have handheld passwords. This is a good security measure and if you make this kind of change after the IT Policy has already been deployed, it will likely aggravate your users who have already become used to not using handheld passwords. It is better to enforce this from day one.

Summary

In this chapter, we have learned about the different components of the BlackBerry Desktop Software and how it has evolved from an all-in-one solution to a very flexible program. We have discussed the different methods of installing and distributing the BlackBerry Desktop Software to all your users, as well as the different options that you have.

We have also learned how to create a silent install, which can help speed up your deployments and keep your desktop configurations constant for all users.

Finally, we learned about the role of IT Policies, as well as how to create these for several versions of the BlackBerry environment.

In Chapter 4, we will cover how you can upgrade your BlackBerry environment. We will discuss how to upgrade from BES 2.2 or 3.6 to BES 4.0. We will offer different methods of upgrading your Lotus Domino and Exchange BES to 4.0.

Upgrading Your BlackBerry Environment

At many organizations, software is not upgraded just because the vendor releases a new version. Usually, the organization will evaluate the benefits of the features of the new version. Once it has been established that there is significant enough benefit to upgrade, a process is put in place to test, pilot, evaluate, and, in many cases, internally certify the software.

When this process is complete and the organization has certified the new software, a plan is created to roll it out to everyone. The BlackBerry software will be no different, and it should not be. It, too, is software that may provide significant benefits to your organization, and it, too, must be tested, evaluated, piloted, and ultimately rolled out. Depending on your organization's IT culture, this process may take up to six months or longer. This chapter should aid you in that process by covering the major topics involved in a BlackBerry upgrade. As always, refer to the manual when performing an upgrade.

Understanding Version Differences

The first step in the upgrade process is to understand the differences between the version that is currently installed and the new version. The decision to upgrade can be as a result of a specific feature or a process that is included in the new version (for example, the multi-domain support in the IBM/Lotus BES). This decision can also be driven by demand from the users.

It is not necessary that every feature be enabled during the upgrade process. It is not unusual to pilot and roll out one feature at a time. The important thing is to design an upgrade process that meets the unique requirements of your organization and takes into account any additional changes you want to make. An example of this is changing the server hardware during the upgrade process.

The following table provides a summary of the major features of the Domino BES 2.2, Exchange BES 3.6, and BES 4.0. (Apart from the mail platform differences, the features of the Domino BES 4.0 and the Exchange BES 4.0 are the same.) For a complete feature list, refer to the documentation supplied with the BlackBerry software.

Feature	Domino BES 2.2	Exchange BES 3.6	BES 4.0 (Domino and Exchange)
Wireless e-mail	✓	✓	✓
Wireless e-mail reconciliation	✓ (No read/ unread flag sync)	✓	✓
Wireless e-mail settings			✓
Wireless calendar	✓ (Some features missing)	✓	✓
Wireless PIM			✓
Mobile Data Service	✓	✓	✓
Wireless IT Policy		✓	✓

Deciding Whether to Upgrade to BES 4.0

As mentioned, the new features and improvements included in BES 4.0 often will drive your decision of whether or not to upgrade your existing system. This section examines some of the benefits offered to users of Microsoft Exchange 3.6 and Lotus Domino 2.2.

Microsoft Exchange BES 4.0

Upgrading your current Exchange BES 3.6 to 4.0 will provide improvements in the area of BES management. BES uses Messaging Application Programming Interface (MAPI) threads to keep track of changes to the BlackBerry users' mailboxes. In general, older versions of the BES allowed one BES to handle about 500 BlackBerry users. To add more flexibility, RIM allowed for the installation of up to four BES instances on one Windows server, which effectively enabled one Windows server to handle around 2,000 BlackBerry users.

When to add the BES instances was a job for the IT staff, and this could become time-consuming. In BES 4.0, there is now only a need for one BES instance per Windows server. If there are other BES instances on the same Windows server at the time of installation, the first BES instance is chosen as the *primary instance* while the others become *secondary instances*. When the BlackBerry user connects a BlackBerry handheld device to a PC and synchronizes, a service book is sent to the handheld unit that moves that user from one of the secondary BES instances to the primary instance. If the handheld device is using BlackBerry 4.0 and running totally wirelessly, this service book is sent OTA.

After all users are moved off of a particular BES instance, a resource kit tool can be used to remove that BES instance and free up the SRP ID. To cope with the MAPI worker thread limitations, the multiple instances of the BES have been replaced by multiple Message Agents, which are spawned automatically by the BES when needed.

BES 4.0 adds new components to the BlackBerry environment that can be installed on separate hardware. This helps to remove load from a single server and also helps the product be more flexible within your environment. These components include the BlackBerry Router, Attachment Service, and the Configuration Database.

Lotus Domino BES 4.0

For Lotus Domino environments, BES 4.0 is a huge step forward. Many new concepts and features help both the end user and the IT staff.

All user configuration data is now stored in a Configuration Database. This database can reside in MSDE or SQL Server. A new management tool called the BlackBerry Manager has a vastly improved interface and set of functions. As with the Exchange BlackBerry environment, new components include the BlackBerry Router and many others that can now be installed on separate servers to reduce load.

Common Upgrade Considerations

With the advent of BES 4.0, the features and components are almost exactly the same on all versions of the BES (Exchange, Domino, and GroupWise). We will discuss the common components, as well as the e-mail environment specific components.

Among your most important considerations are the following:

- ❑   Deployment of SQL or MSDE
- ❑   State databases
- ❑   Attachment Service
- ❑   BlackBerry Router
- ❑   MDS

Deployment of SQL or MSDE

If you are planning on upgrading to Domino BES 4.0, the deployment of SQL or MSDE is a brand new concept to you. You should carefully consider your options when planning this upgrade.

With Domino BES 2.2 or earlier, all user configuration information (including the user's Domino name, location of the user's mail file, encryption key, signature, and mail filters) was stored in a Domino database called `BlackBerry Profiles`. When you upgrade to BES 4.0, the contents of the BlackBerry Profiles database will be migrated to a new location. This new location will either be a local MSDE database or a full SQL database housed on an instance of SQL Server.

If you are upgrading from Microsoft Exchange BES 3.6 or earlier, you will already have made this choice some time ago, based either on your desire to use the MMC snap-in, or later when RIM stopped supporting the proprietary BlackBerry Management application. When you started using the MMC snap-in, you were required to use either MSDE or SQL Server. During the upgrade to BES 4.0, you may choose to leave your SQL Server or MSDE implementations intact, or you may want to reorganize it to make it more flexible or better suited for disaster recovery.

No matter which version of BES 4.0 you are upgrading to (Microsoft Exchange or Lotus Domino), you should consider the following factors:

- ❑ MSDE
- ❑ SQL Server
- ❑ BES and SQL Replication

Microsoft Database Engine (MSDE)

Microsoft SQL 2000 Database Engine (MSDE 2000) is Microsoft's free database engine, which is included in the BES installer. This database type is ideal for small organizations or single BES server installations. MSDE does have some limitations, however. Because this is the free version of Microsoft SQL Server relational database, it cannot scale as well as an SQL Server can. It also has a Workload Governor, the function of which is to introduce an artificial delay when the number of concurrent connections exceeds eight. Detailed information can be found on the Microsoft Web site (search for "MSDE 2000").

Additionally, MSDE 2000 does not come with any management tools (although some third-party tools are available), nor does it support replication between database instances.

SQL

SQL 2000 is the full version of MSDE 2000 and is the preferred choice for large installations with large numbers of users. If your organization already uses SQL Server for other applications, consider using this infrastructure for your BES Configuration Database. Your Database Administrator (who may have already deployed applications that rely on the SQL Server infrastructure) has most likely already taken into account topics such as disaster recovery and high availability. (See Chapter 8 for more information on disaster-recovery plans.)

BES and SQL Replication

BES 4.0 does not support two-way SQL replication. In a multi-office environment connected through WAN links, you may decide to point all BES servers at a single SQL database in a particular office with all administration performed on this single SQL database via the BlackBerry Management tool. You may also decide to deploy a single SQL Server or MSDE in each office. Depending on the number of users the BES is processing, you may even install the SQL server on the same Windows server as your BES.

You could implement a hybrid of these two scenarios by having a regional SQL server and having all BES servers in a particular region point to that server. The regions could be broken up by country, continent, or something in between, depending on the speed of the WAN links between each location and their available bandwidth.

Because MSDE does not support replication, if you plan to have a disaster-recovery plan that calls for a replica of the database to reside on a second server in another location, you will need to use SQL.

If your disaster-recovery plans call for a simple in-place restore onto new hardware, you can then use MSDE, since replication is not required. If your organization is small or only has one BES, you can use MSDE.

State Databases (Domino BES Only)

The *state databases* are unique to the Domino BES because they provide the only mechanism by which to track messages that are processed by the BES. They also provide a link between the messages on the handheld device and the messages in the user's mail file.

In BES 4.0, the state database takes on the role of a personal outgoing queue. In BES 2.2, one outgoing queue is used for all users on a particular BES. In BES 4.0, the common outgoing queue is gone, and the functionality has been incorporated into each user's state database.

State databases become even more important when planning for a disaster. If your disaster-recovery plan calls for an in-place restore of all server data, then you do not have to give much careful consideration to state databases. If the primary server goes down, you will simply restore the data onto a new server with the same name and start it up.

However, if your disaster-recovery plan calls for a more instant recovery, you will need to consider seriously the state databases. In BES 4.0, you can seamlessly move users from one live BES to another. When you perform this move, the user's state database is replicated from the source BES to the destination BES, and then removed from the source. To system users, nothing has happened, and they continue to work normally.

If the source BES is down, the user cannot be moved, because the state database cannot be found. If a disaster occurs and you want to move users from one BES to another, you can only do so if the primary BES is up and running. Typically, this is not the case, and the primary BES would be unavailable.

To plan for this scenario, you must decide in advance which users will be moved to which BES. When you know how the users will be moved between the BlackBerry Enterprise Servers, you must set up one-way replication of those state databases within Domino. With a current replica of the users' state databases on the destination BES, a seamless move can take place. Ultimately, this would mean that both BlackBerry Enterprise Servers must be live at all times, and so extra licenses must be purchased.

Using this method, you could choose to consolidate your BES users onto a single BES in case of a disaster, or purchase a backup server for each BES.

Attachment Service

This service is not new to either Domino or Exchange BES environments, but in BES 4.0, it has added functionality. This added functionality allows BlackBerry users to view more types of attachments on their handheld devices, including different types of image files. This could cause the service to begin using up more CPU time on your BES, and so you may want to install it onto a dedicated server or a server that can accommodate it.

BlackBerry Router

This new service in BES 4.0 functions as a broker between different communication media. It ensures that the user's handheld device is always communicating with the BES using the fastest possible medium. For example, if the handheld device is on the cellular network and the user plugs it into a computer using a USB or serial cable, the BlackBerry Router will begin routing data between the handheld unit and the BES over the LAN instead of over the cellular network.

One thing to remember is that any data traffic that does not use MDS will still go through the wireless network. This includes Wireless Access Protocol (WAP) browser traffic, Internet browser requests to the BlackBerry Internet Browser Service (BIBS), Web mail, or any other handheld application that does not use MDS.

The BlackBerry Router also handles all communications between the BlackBerry NOC and the BES.

Your security team may insist on having the BlackBerry Router in the DMZ, as previously discussed in Chapter 2. You must remember that if it is in the DMZ, it may be limited from functioning correctly unless it can communicate with all handheld units on port 4104, and with the BlackBerry NOC on port 3101. This fact alone may make it an undesirable candidate for the DMZ, and it may turn out better to install it behind the firewall.

MDS

MDS is not new in BES 4.0, but it does provide added functionality in the area of application development. This fact, coupled with the improved BlackBerry Browser (which uses MDS), may mean an increased use of MDS in your organization.

If you think that MDS usage will increase (or has already increased) to such an extent that the MDS service is using up too many CPU cycles and causing the main BES task to slow down, you may want to move it to its own dedicated server. You could install MDS on the same server as the Attachment Service.

Preparing to Upgrade

In preparation for upgrading your BlackBerry environment, there are many considerations. These include which patch levels your BES must be at before the upgrade, what maintenance you must perform before upgrading, and in what order to upgrade (BES first or handheld devices first).

Patch Levels

If you are running a Lotus Domino BES, prior to upgrading, you must be running BES 2.2, Service Pack (SP) 1 or later. If you are still running BES 2.1, you must upgrade to BES 2.2, SP3.

If you are running a Microsoft Exchange BES, prior to upgrading, you must be running BES 2.1, SP5; BES 3.5, SP2; or BES 3.6, SP3 or later.

Maintenance

Before you actually upgrade your BES, you should consider performing maintenance on your Domino, MSDE, or SQL Server databases.

For a Domino BES, ensure that you run `compact`, `fixup`, and `updall` on all databases within the BES directory structure. These include the BlackBerry Profiles (`blackberryprofiles.nsf`), BlackBerry Directory (`bbdir.nsf`), outgoing queue, and all state databases. This maintenance is to ensure that the databases are healthy and without errors.

In addition to running maintenance on the databases, you should inspect them to ensure that they do not have any save or replication conflicts. If you find save or replication conflicts, you should shut down the BES task and delete the conflicts.

In a Microsoft Exchange environment, if you are using MSDE 7, you must manually upgrade it to MSDE 2000 because the BES installer will not do this for you.

Upgrade Order

When the BlackBerry 4.0 environment was conceived, RIM devoted considerable attention to the upgrade path to ensure that you have the most choices possible for different upgrade scenarios. As a result, RIM developed the upgrade process so that all scenarios provide backward compatibility with each other, which allows you to upgrade to the version 4.0 platform in any order you choose.

Upgrading the BES First

If you choose to upgrade your BES server(s) to BlackBerry 4.0 first, your users will not notice any change in the BlackBerry experience. To them, there will have been no upgrade and their Desktop Manager and handheld device will continue to function as they always have.

In the Lotus Domino environment (specifically, where the Desktop Manager communicates with the `blackberryprofiles.nsf` database), since Domino BES 4.0 keeps this database in sync with the SQL Server or MSDE database, the desktop experience and support mechanisms will remain the same.

As time permits, and using the method you choose to employ, you can upgrade the users' handheld units. As each user receives an upgraded handheld device, he or she will see (and be able to take advantage of) the new features of the BlackBerry 4.0 environment.

Upgrading the Handheld Devices First

If you choose to upgrade your handheld devices to BlackBerry 4.0 first, your users will see some cosmetic and interface changes on their handheld units. However, they will not be able to use any of the new BlackBerry 4.0 features until you upgrade their BES.

The BlackBerry 4.0 Handheld Software is designed to be backward compatible with older versions of the BES, so there should be no incompatibilities to worry about. Of course, your users may start asking questions about the new features that they see on their handheld devices (such as the upgraded browser, the Password Keeper, and the Pictures feature, to name a few). This may put pressure on you to move the BES upgrades forward more quickly than you planned, so that your users can make use of the new features.

Upgrading Your BES

While many of the upgrade scenarios and types of upgrades are common among all types of BES systems, there are some differences between upgrading an Exchange BES and a Domino BES. This section first examines the differences between the Exchange BES upgrade and the Domino BES upgrade. We then turn our attention to common BES upgrade scenarios.

Upgrading Your Exchange BES

When you upgrade an existing Exchange BES, a number of operations are performed. The following list is a summary of the tasks that the BES installer performs:

1. The installer replaces all the BlackBerry system files with the newer 4.0 version.

2. The installer removes all non-relevant BlackBerry registry entries.

3. The MSDE or SQL Server database schema is updated and the data is migrated.

4. Last, and most important, all BES instances are collapsed into a single instance, with a single SRP ID and authentication key. This does not occur instantaneously, but takes a while to complete depending on how quickly handheld devices receive their new service book.

After the BES installer completes and your BES is running version 4.0, the following steps occur as time passes:

1. During the upgrade, the first instance that was installed on the server becomes the primary instance. The remainder of the instances will become secondary instances. All new users added to the BES will be assigned to the primary instance.

2. As existing users cradle their handheld devices to synchronize their user data, a new service book containing the new server information (that is, the primary instance SRP ID) is sent to the device. Existing users on the primary instance continue as usual.

3. After all the users have been migrated to the primary instance, the other secondary instances can be removed using the BlackBerry Resource Kit, which can be downloaded from the `blackberry.com` Web site.

Remember that the additional instance SRP IDs are not full-fledged SRP IDs. Therefore, those SRP IDs must be converted to full SRP IDs before they can be reused on other BES servers.

Upgrading Your Domino BES

When you upgrade an existing Domino BES, a number of operations must be performed. The following list is a summary of the tasks that the BES installer performs:

1. The installer replaces all the BlackBerry system files with the newer 4.0 version.

2. The installer removes all non-relevant BlackBerry registry entries.

3. MSDE is installed, a new database and schema is created; or, the installer connects to the SQL Server you have specified, creates the new database and schema.

4. The installer migrates all configuration data from the `blackberryprofiles.nsf` Domino database to the new MSDE or SQL Server Configuration Database.

BES Upgrade Scenarios

There are two different approaches to the upgrade process. The first is an in-place upgrade, which implies that you do not change the hardware that your BES is running on but simply upgrade the BES software to 4.0. The second is a migration upgrade, which implies that you build a new, more powerful server, cut-over to it, and do an in-place upgrade. This second scenario may turn out to be your best option if you feel that your current BES server will not be able to handle the extra load placed on it by the new BES 4.0 features.

In-Place Upgrade

Before you decide to upgrade your existing BES server using the same hardware, you must be aware that the new features and additional services (as well as the change in user patterns) will place an additional load on your BES server. If you suspect that your server hardware is not adequate, refer to your baseline data and compare that with the current load on the server. This will provide you with an accurate picture of the usage of your server, and how much spare capacity you have on your current server hardware.

If your server is operating at fairly close to maximum, or if you suspect that the new features and user patterns will push you beyond the limit of your existing hardware, you could first try moving the Attachment Service to a different server. If your users read several attachments on their handheld devices, this service move may be all that is required. If you have already done this, you will have to consider purchasing new hardware for your BES server. If you are in the situation such that purchasing new hardware will unduly delay your BlackBerry upgrade project, consider upgrading your BlackBerry server, but delay enabling the new features as long as possible.

Additionally, BES 4.0 requires much more disk space for log files. The number of log files has dramatically increased, so ensure that your server has sufficient disk space. In addition to the increased disk space requirements, and since there is so much more logging, it is far better to configure the BES to write its log files to a disk other than the disk on which the BES is installed. You could even place the log files on a separate server.

Clean Up User Accounts

Before you upgrade your BES, check for users who have BlackBerry accounts but do not have e-mail accounts (such as users who have left the organization). These unused accounts will unnecessarily consume valuable system resources.

Check the pending message counts for your users and identify the users who have very high pending message counts. These users most likely have either left the company, or they do not use the BlackBerry.

For upgrades to the Exchange BES, be sure to run `handheld cleanup` on the BES server to refresh the Server and User information.

Clean Up Old Log Files

Check your server for old log files. You may be surprised to find that your log files have grown to several gigabytes in size. On the upgraded server, consider limiting the length of time that you will keep the log files. A minimum of 7 days is required for troubleshooting purposes, but it is recommended that log files be kept for 30 days.

You are now ready to run the BES 4.0 installer on your server to upgrade it.

Migration to New Hardware

This upgrade scenario makes use of the *knife-edge cut-over* disaster-recovery scenario and the in-place upgrade. The knife-edge cut-over is the process whereby you shut down the production BlackBerry server and switch on the secondary BlackBerry server. The secondary server is an already installed and configured BlackBerry server (which is essentially a standby server). This procedure is usually used in high-availability scenarios where very little or no downtime is allowed.

Create the Knife-Edge Cut-Over Server for Exchange Systems

For the purposes of this scenario, we will assume that you are using BES 3.6. Follow these steps:

1. On a server that already has the operating system installed and configured, install the BlackBerry server software. It is recommended that this procedure be done during a maintenance window or scheduled outage period, so that the production server can be turned off.

2. Disconnect the production server, or, better yet, shut down the production server. The reason for this is to prevent the same SRP ID from connecting to the BlackBerry NOC at the same time. If this happens, a lockout condition will occur and the SRP ID in question will be disabled. This is done to prevent someone spoofing your SRP ID, to prevent denial-of-service attacks on the BlackBerry NOC, and, most important, to ensure that e-mail is delivered to the correct server.

3. When the production server is down, configure the standby server as follows:

    a. Use the same BlackBerry server name.

    b. Connect the BlackBerry server to the same configuration database, if this database is on a separate server. If you do not have your server configured to connect to a separate Configuration Database, you will not have access to your server's MDS settings, IT Policy information, or license keys. All this information will have to be maintained separately.

    c. Bring up the BlackBerry server and ensure that e-mail does, in fact, flow to and from the BlackBerry handheld devices.

4. You may have to make certain network changes, depending on how you intend to activate the standby server. You could write a script to change the identity of the standby server to be the same as the production (that is, same NT Name, same IP address, and so on), or you could modify your DNS make the standby server an alias to the production server. You should develop a strategy that suits your needs and test this in a lab environment.

This process works because the BES 3.6 stores its information in hidden folders in the BlackBerry Service account's mailbox. The standby server will connect to the same Service account mailbox as the production server and, therefore, have access to the same information. User-specific information is stored in hidden folders in the user's mailbox.

Create the Knife-Edge Cut-Over Server for Domino Systems

For the purposes of this scenario, we will assume that you are using BES 2.2. Follow these steps:

1. On a server that already has the operating system and Domino server installed and configured, install the BlackBerry server software. It is recommended that this procedure be done during a maintenance window or scheduled outage period, so that the production server can be turned off.

2. Stop the Domino server (but leave Windows running). The reason for this is to prevent the same SRP ID from connecting to the BlackBerry NOC at the same time. If this happens, a lockout condition will occur and the SRP ID in question will be disabled. This is done to prevent someone spoofing your SRP ID, to prevent denial-of-service attacks on the BlackBerry NOC, and, most important, to ensure that e-mail is delivered to the correct server.

3. After the production Domino server has been shut down, install the BES software on the standby server and ensure that it is at the same service pack level as the production server. Use the same SRP ID and key during the installation process. When you have completed the BES install, edit the notes.ini file and comment out the BES task so that it does not start automatically.

4. Start the Domino server, start the BES task by typing **load bes** at the Domino server console, and allow it to initialize correctly. Then, shut down the Domino server (leaving Windows running).

5. Now you need to transfer over all Domino databases that reside in the \lotus\domino\data\bes subdirectory. You can do this by simply doing a file copy of the entire BES directory.

6. Once all of the files have copied over, start the Domino server and BES task by typing **load bes** at the Domino server console. Ensure that you can send and receive e-mails on your BlackBerry system.

Perform the Knife-Edge Cut-Over and Upgrade

To migrate to new hardware, you first configure the new server as a standby server, running the same code as your production server. Then, during your upgrade window, you will turn off the production server, do a knife-edge cut-over to the new hardware and verify that it is working correctly. You can then do an in-place upgrade, as described previously.

Summary

This chapter has discussed the various factors involved with the BES upgrade process. We examined whether the new BlackBerry 4.0 features would drive your need to upgrade to version 4.0.

We discussed the options of whether to upgrade your BES servers or your handheld devices first. We examined how, no matter which one you choose to do first, they are backward compatible with each other. We also discussed the many factors that you need to take into consideration when planning an upgrade, including whether to use MSDE or SQL Server and where to install the different BlackBerry components.

Finally, we discussed some options for upgrading your BES servers, including an in-place upgrade and a migration upgrade.

In Chapter 5, we will discuss upgrading the BlackBerry handheld devices. We will cover topics such as upgrading in a 4.0 BES and pre-4.0 BES environment. We will discuss how upgrading the handheld devices has become easier to administer and execute using the new functionality of BES 4.0 Software Configurations in conjunction with the Handheld Manager.

Installing or Upgrading the Handheld Software

You need to upgrade the software on your users' handheld devices for two reasons: to correct any potential bugs that may be discovered in the Handheld Software that might affect the functionality of the handheld unit, and to add new features to the handheld device.

When upgrading your BES from version 2.2 or 3.6 to 4.0, your users will not benefit from the new features unless you upgrade the software on the handheld device. This chapter discusses this upgrade procedure for older and current BES installations. The Handheld Software contains the BlackBerry operating system, Virtual Machine (VM), and all core applications (such as e-mail, calendar, address book, browser, and more).

Upgrading in a Pre–BES 4.0 Environment

In a pre-4.0 BlackBerry environment, your only option for upgrading the software on the handheld device is to install it on each user's computer and use the Desktop Manager.

Using this process, you would double-click the Application Loader icon in the Desktop Manager and navigate through the screens to upgrade the Handheld Software.

Upgrading in a BES 4.0 Environment

If you are planning to upgrade your BlackBerry environment to version 4.0, or you are already in an all-4.0 BlackBerry environment, then you have a few more options. If you still use the Desktop Manager on all of your user's computers, then you could opt for continuing to install the software for each handheld device on each computer, and then use the Desktop Manager to upgrade the handheld devices.

If you have chosen to remove the Desktop Manager as part of your BlackBerry 4.0 migration plans, or you have already removed the Desktop Manager and now rely solely on the Handheld Manager, then you can benefit from the new upgrade methods by removing manual reliance on the Desktop Manager in the upgrade process.

The upgrade basically consists of the following steps:

1. Create a shared server.
2. Create software configurations and assign them to the handheld units.
3. Upgrade or install the software on the handheld units.

Creating a Shared Server

The new Management Console features of BES 4.0 enable you to create software configurations for the handheld units. These software configurations make use of a read-only share on a Windows server located on your network.

To use the software configurations, first choose a Windows server on your network that has available space and is not too busy. You may want to build a new Windows server for this task, or use an existing Windows server that may be hosting the Attachment Service or BlackBerry Router. If you have a small installation and your BES server is not busy, you can use it to host these files.

When you have a Windows server ready, install all the Handheld Software that you are going to need for the handheld units. You can either use the installation software on the CD-ROM provided with a BlackBerry wireless device, or download the software from your carrier's site. Handheld Software version 3.8 and later does not require that the Desktop Manager be installed first.

When you install the device software, it will install to the c:\program files\common files\ research in motion directory. If you browse to this directory, you will see a few subdirectories. The apploader subdirectory is where the Application Loader is installed. The shared subdirectory is where the actual device software is installed. Under the shared subdirectory, you will find another subdirectory called loader files. Under loader files, you will see further subdirectories that contain the actual Handheld Software for each device.

Before continuing, you must share the directory c:\program files\common files\research in motion as read-only to the group Everyone. If you are not using domains and only have a workgroup, you must enable the Guest account and allow it read-only access to this directory.

The software configurations contain Handheld Software. However, they can also contain third-party applications. To add third-party applications to your software configurations, create a new subdirectory called applications under the shared subdirectory. Under the applications subdirectory, create new subdirectories for each software application or group of applications that you want installed on the devices. For example, you could create a directory called color games and bw games, or even break it down by software vendor. Then, copy over the .alx, .cod, and .dll files from each third-party application into the respective subdirectories.

You must preserve the directory structure of the original application when you copy over the files. To ensure that you are preserving the directory structure, install the third-party application on a PC and copy over the directory structure and relevant files (remember to copy over only `.alx`, `.cod`, and `.dll` files). Here is an example of what your directory structure might look like:

```
C:\Program Files\Common Files\Research In Motion\Shared\Applications\MagMic Color
C:\Program Files\Common Files\Research In Motion\Shared\Applications\MobileSSH BW
C:\Program Files\Common Files\Research In Motion\Shared\Applications\MobileSSH
Color
```

Any time you make changes to the files in the `c:\program files\common files\research in motion\shared` directory, you must instruct the Application Loader to re-index to include the changes within the index. The Application Loader keeps an index of all installed Handheld Software and all third-party applications that can be included in Software Configurations. To update its index manually after you add new software or handheld code, open a command window, and type the following:

```
cd c:\program files\common files\research in motion\apploader
```

Then, type the following and press Enter:

```
loader/index
```

After a few seconds, the index will be complete and the next time you edit a software configuration, you will see the new applications.

Creating Software Configurations

After you have your shared server in place, you can create handheld software configurations. These software configurations consist of the Handheld Software and third-party software that you want to deploy to the devices, plus a policy that controls the behavior of the third-party applications after they are installed. If you are in a Domino BES 4.0 environment, open your BlackBerry Manager console (see Figure 5-1). If you are in an Exchange BES 4.0 environment, open your Handheld Manager console. The functionality and look and feel of the Exchange BES 4.0 Handheld Manager is not included in the Exchange BES 4.0 BlackBerry Manager.

Notice that the screen is divided into three main parts. On the left side of the screen are listed such items as the BlackBerry Domain and Handheld Management. Clicking the plus sign to the left of these entries expands the view to show subentries. In the top half of the right side of the screen, you will find the installed software configurations. In the lower half of the right side of the screen, you will find the following tasks:

- ❑ Add New Configuration
- ❑ Edit Configuration
- ❑ Copy Configuration
- ❑ Delete Configuration
- ❑ Manage Application Policies

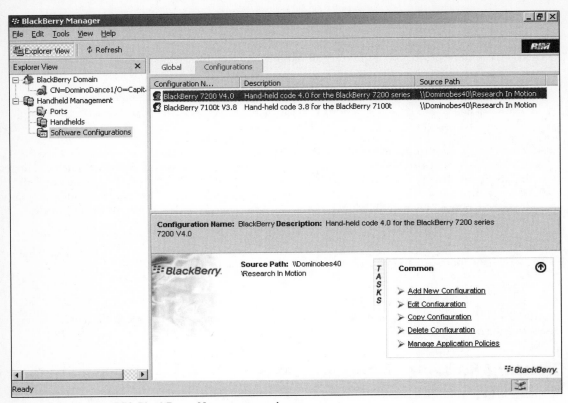

Figure 5-1: Domino BES BlackBerry Manager console

Obviously, the first four tasks in this list enable you to work with (that is, add, edit, copy, or delete) configurations. The last task in the list enables you to manage policies for your applications.

Working with Configurations

Click Software Configurations below Handheld Management. In the bottom right-hand side of the screen, click Add New Configuration.

The next screen to display will be the Software Configuration page (see Figure 5-2). In the first field on the page, type a name for the configuration. Enter a description for the configuration in the next field. For the Handheld Software Location field, you can either type in the path to the network share or select Change to browse for the location of the share. Note that this field must contain a Universal Naming Convention (UNC) name and cannot be a drive mapping.

After you have typed in the UNC to the shared volume, you will see all of the handheld and third-party software. Simply check the boxes next to the Handheld Software and third-party applications that you want for this particular software configuration.

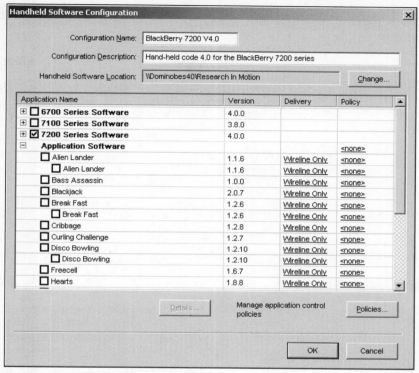

Figure 5-2: Software Configuration page

On the right-hand side of the main pane in the Software Configuration screen, you will notice the following two columns:

❑ **Delivery:** This column enables you to choose the network on which the third-party applications will be installed. For the device software, the delivery method is permanently set to Wireline Only. This makes sense, because it is not yet possible to install the software for the device over the air. However, for third-party applications, you can choose to have the applications installed over the LAN only (Wireline Only) or over the air (Wireless).

❑ **Policy:** The Policy column indicates the policy that is assigned to the particular third-party application. The handheld software cannot have a policy assigned to it and so the policy field will remain set to None. Next to each third-party application, you will be able to change the policy. The policy governs how the third-party application behaves on the device, if it is allowed to be installed or not, and how it behaves with the firewall for the device.

Managing Application Control Policies

By default, there are no application policies, and so you must create application policies first. From the main Software Configuration screen you can click the Manage Application Policies task. If you are working from a Software Configuration screen, you can also click the Policies button in the lower right-hand portion of the screen. After you have invoked the task by either method, you see the screen shown in Figure 5-3. You can use this screen to create and edit your application policies.

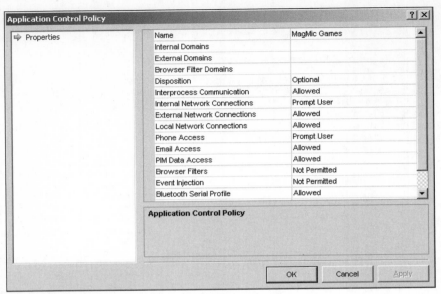

Figure 5-3: Application Control Policy screen

If you choose not to create any application policies, the third-party applications will behave in the Default manner on the device.

Assigning the Software Configurations

With your software configurations created, you must now assign them to your handheld devices. From the left-hand pane of the BlackBerry Manager screen, shown in Figure 5-1, click the Handhelds option listed under the expanded Handheld Management entry. A list of the handheld units that are registered on this BES will be displayed, as shown in Figure 5-4. If the device is already assigned to a user, the user's name will be displayed.

If you have a device that is not yet assigned to a user, you can assign it. To do this, click the Handhelds view, click the user who does not yet have a device assigned, and then click Assign Handheld to User. Choose from the available, connected devices.

Choose one or many devices and click Assign Software Configuration from the list of tasks shown in the lower right-hand portion of the screen. The next screen will list your software configurations, as shown in Figure 5-5. Choose one and click OK.

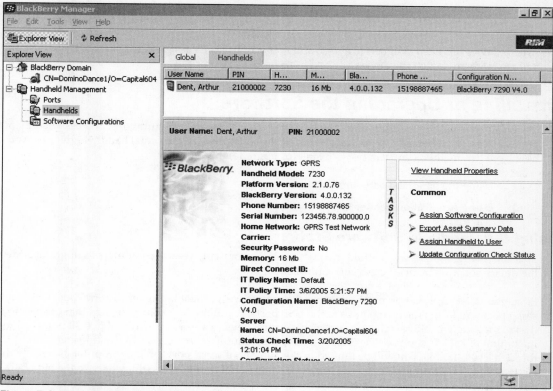

Figure 5-4: Users assigned to handheld devices

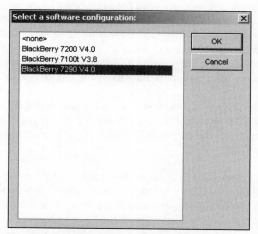

Figure 5-5: Assigning a software configuration
to a handheld device

With the software configurations now assigned to your devices, every time the Application Loader is executed from the shared volume, the Application Loader will know exactly what to install on the device.

After a device has been assigned to a user, and that user has been assigned to a Software Configuration, you are ready to deploy the device.

Installing or Upgrading the Software

Once you have your shared server in place, created your software configurations, and assigned them to your devices, you are ready to upgrade or install the software onto the actual handheld devices. You now have two methods you can use to install (or upgrade) the software:

❑   Using the handheld software station

❑   Using Application Loader Lite

Using the Handheld Software Station

The *handheld software station* method involves using one or many central computers to upgrade or deploy the device software in an efficient and controlled manner.

First, you must have one (or many) desktop computers with available Universal Serial Bus (USB) and serial ports. You can attach USB hubs to the USB ports. If you use BlackBerry devices that connect to the desktop through a USB cable, you can attach as many devices as you have available USB ports (up to a maximum of 256). If you use BlackBerry devices that require a connection to a desktop computer through a serial cable, you will be limited to the number of serial ports on the computer. Normally there are two serial ports, although you can purchase multi-serial port devices (think back to the old UNIX days) that will add many serial ports to one computer. These desktop computers, when connected to the BlackBerry device, now will serve as *software station computers*.

On each of the software station computers, install the BlackBerry Manager or Handheld Manager console and configure it to connect to your SQL database. When you connect to the Configuration Database, the Software Configurations will already be there, so there is no need to re-create them. However, if you have not yet set them up, remember not to map a drive letter to the shared volume, but rather use a UNC entry. If you map the share to a drive letter, this process will not work. See the section "Software Configurations" earlier in this chapter for more information on how to create and assign Software Configurations.

You can use the handheld software station method to upgrade or install any device that must run Handheld Software version 3.8 or later. Handheld Software below version 3.8 cannot be installed on the shared server, because it requires the presence of the Desktop Manager. Any handheld can be installed or upgraded using this method.

After you have connected a device (or devices) to the computer, return to the BlackBerry Manager screen. From the left-hand pane of the BlackBerry Manager screen, shown in Figure 5-6, click the Ports option listed under the expanded Handheld Management entry. You will see the ports used by the handheld units listed in the top half of the right-side of the screen in the Ports view.

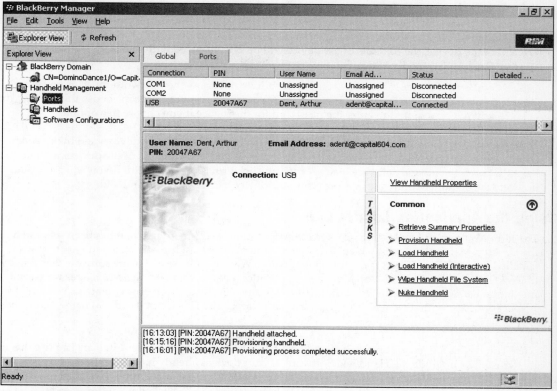

Figure 5-6: Ports view

When you click on an individual device, you will be able to perform the following actions (or tasks) listed on the right-hand side of the lower half of the screen:

❑ **View Handheld Properties:** Enables you to view the handheld properties, including what device hardware is being used, what state the battery is in, how much memory is left, and what software is being run on the device (including all third-party applications).

❑ **Provision Handheld:** Provisioning a handheld involves sending certain Service Books to the device that configures it to communicate with the BES correctly. After the Service Books have been sent, the user's data is synchronized with the device and any third-party software programs are installed.

❑ **Load Handheld:** Loading the handheld will load the operating system and any third-party applications that are assigned to the user.

❑ **Load Handheld (Interactive):** This is the same as the Load Handheld option. However, it provides the capability to make changes that override the settings in the Software Configuration.

❑ **Wipe Handheld File System:** If you connect to the software station computer a device that previously belonged to someone other than who you want to provision it for, you must wipe (delete) the file system on the device before you provision it. If you do not wipe the file system of the device, the contents of the device will be synchronized with the user's network data. If the device is totally blank, you will be able to provision it safely.

❑ **Nuke Handheld:** If you receive a device, but do not know the password for it, or you simply want to destroy all data and handheld software on the device, you can choose this option. You will not be required to enter the password for the device if you choose this option.

To deploy one or many devices, click on the Ports view. Select one or many devices and make your choice using the Task list at the lower half of the screen. If you know that all of the devices are blank, but have the correct handheld software loaded, you can click Provision Handheld. If you want to install or upgrade the Handheld Software and provision the device, click on Load Handheld.

Using the Application Loader Lite

The term Application Loader Lite refers to the Application Loader software that has been separated from the Desktop Manager. In pre-4.0 BlackBerry environments, the Desktop Manager software that was installed on the PC contained the Application Loader. With the release of 4.0, the software was renamed to Desktop Software. The Desktop Manager, Application Loader, and Handheld Manager were separated into distinct components. In addition, the Handheld Software itself now contains a version of the Application Loader that we refer to as Application Loader Lite. This is not its official name, but rather a way of describing its capabilities.

Since you have a shared volume that already contains the Application Loader Lite, you can use this utility to install or upgrade the Handheld Software. If your devices already have BlackBerry 4.0 installed, but you still use the Desktop Manager to synchronize with a special application, you must use this method of upgrading, since it is the only way that the device can be backed up and restored.

To do this, you must have access to this shared volume through a UNC path. When you are connected to it, you must run the Application Loader Lite utility. In Windows Explorer, browse to the UNC share.

```
\\<shared server>\Research In Motion\
```

To run Application Loader Lite, change to the Loader directory and double-click `loader.exe`.

In the screen that appears, choose the device that you want to upgrade and proceed. Be sure to back up the software on the device before you upgrade it.

Because a software configuration has already been assigned to the device, you should simply be able to accept all of the defaults. After the device has been upgraded, the Application Loader will restore the data on the device.

This method of upgrading a device from a previous version to version 4.0 can be done by sending a technician around to each user's computer, or by sending an e-mail to all of your users instructing them on how to run the Application Loader. Of course, you must feel comfortable that your users have the patience and knowledge to do this. Otherwise, you should have them send their BlackBerry devices to the technician to be processed.

Upgrading a 4.0 Handheld

If your handhelds are already running version 4.0, doing a full wireless synchronization, and backing up software through a wireless connection, you can upgrade the devices using the software station method with much more efficiency. The reason that this process is more efficient is because there is no need to back up the device data — the data will be synchronized after it has been upgraded. Any user-specific items on the handheld (such as positions of icons on the home screen, auto-text entries, browser bookmarks, and so on) will be restored to the device over the air after the upgrade.

To do this, connect as many devices as you like to the software station computers. Run the BlackBerry Manager or Handheld Manager console and click on handheld units. Then select one or many devices of the same type (say all 7200 series with the exception of the 7290s, or all 6700 series) and click Load Handheld.

The BlackBerry Manager or Handheld Manager will proceed to upgrade the Handheld Software on all of the devices. The wirelessly backed-up data, the last three days of e-mail (depending how you have configured this), and all PIM data will trickle down to the device over the next few hours. If it is connected to a computer running the Handheld Manager, this process will be quicker.

Summary

This chapter has discussed how to upgrade the devices for your users in a pre-4.0 and post-4.0 BES environment. Those methods include using the Desktop Manager with the Handheld Software installed on each computer, using the handheld software station method (where you can connect multiple devices to a single computer and upgrade them all at once), and using a shared copy of the Application Loader Lite.

In Chapter 6, we will cover how to monitor your BlackBerry environment after it has been installed. We will discuss how to set up alerts for system failures, how to set up monitoring using the provided BlackBerry components and third-party Simple Network Management Protocol (SNMP) products, and how to know when to upgrade or load balance your BlackBerry environment as it grows.

Monitoring and Enhancing Your BlackBerry Environment

Adding BlackBerry devices has an impact on your network environment. It is important to gauge how much of an impact this has initially, and how that impact may increase over time, as more e-mail users become BlackBerry users.

In addition to monitoring your mail servers, it is important to monitor your BES servers because they, too, can become overworked. This chapter discusses the effects of BlackBerry users on your mail servers, and which performance indicators you should watch. This ensures that you will be able to upgrade your mail servers and BES servers (or offload users from them) to keep them running at an optimal performance level.

This chapter begins with a look at some tools you can use to monitor performance levels in your BlackBerry environment. Then we will examine just what you should be watching.

Monitoring Your BlackBerry Environment

You can use a number of tools to monitor your BES and mail servers. RIM supplies a monitoring tool called *BlackBerry Enterprise Server Alert* (*BESAlert*) that is a Windows NT service running on your BES server and monitoring the NT Event Log. You can configure BESAlert to send you e-mails when it detects certain kinds of events. You can also use *Simple Network Management Protocol* (*SNMP*) to monitor your servers. RIM supplies the SNMP *Management Information Database* (*MIB*) for the BES on the installation CD-ROM. BESAlert is available for all versions of BES. In addition, the BES SNMP MIB is supplied on the CD-ROM of all BES versions.

It is also important to use the *Performance Monitor* (*PerfMon*) to monitor certain critical indicators on your mail servers and BES servers. You should also be familiar with the monitoring capabilities of the *Mobile Data Service* (*MDS*). When things go wrong, having a good understanding of the BES log files always helps.

Using BESAlert

BESAlert can be configured to monitor the Windows Event Log and watch for alerts that are related to the BlackBerry components. You can configure BESAlert to send alerts based on the alert level, so that you are not overwhelmed with e-mail traffic about your BlackBerry server.

For example, you can configure BESAlert to e-mail you when it sees Critical, Error, or Warning alerts. These kinds of alerts normally mean that a problem has occurred and, based on the text of the alert, you can take the appropriate actions to correct it. You probably will not want to configure it to send e-mail to you when it sees Informational alerts, because they do not indicate a problem and there are many Informational alerts logged all the time.

You may need to configure BESAlert to send e-mails to a group, rather than just yourself. It is always better if a group of people is notified of a problem. For example, if the alert is telling you that the BES has lost connection to RIM, then it is unlikely that you will receive the e-mail if you are away from your desk and have only your BlackBerry device with you.

Pre-4.0 BES

To configure BESAlert in a pre-4.0 BES environment, you must run the *BlackBerry Enterprise Server Monitor (BESMonitor)* program. The BESMonitor program is a small monitoring program that reports the status of one or many BES servers. It will tell you visually through the use of system tray icons whether the BES server or servers that it is monitoring are connected, down, or not connected. In addition, BESMonitor will display the statistics shown in the following table.

Statistic	Description
BESFromHH	Indicates the total number of messages that were sent from the handheld units in this session.
BESIncoming	Indicates the total number of messages that were received by the BES from handheld units.
BESOutgoing	Indicates the total number of messages that were sent to BlackBerry handheld units by the BES.
BESToHH	Indicates the total number of messages that were sent to handheld units in this session.
SRP	Indicates whether the BES has an active connection to the wireless network.

To run the BESMonitor, browse to the icon in the Start menu of your BES server. When BESMonitor is running and you have configured it to connect to a particular BES, click the Configure button. This brings you to the BESAlert screen shown in Figure 6-1.

Here you configure BESAlert to send an e-mail to a particular e-mail address through a particular SMTP server when a particular Event Log entry is seen.

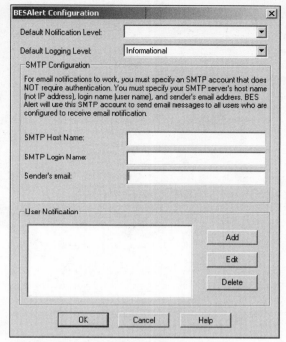

Figure 6-1: BESAlert Configuration screen

BES 4.0

To configure BESAlert in a BES 4.0 environment, you must use the BlackBerry Manager tool. Find the BlackBerry Manager icon on the desktop of your BES and double-click it.

When it is open, click your server name on the left-hand side of the screen shown in Figure 6-2. On the right of the screen, click Edit Properties on the Server Configuration tab.

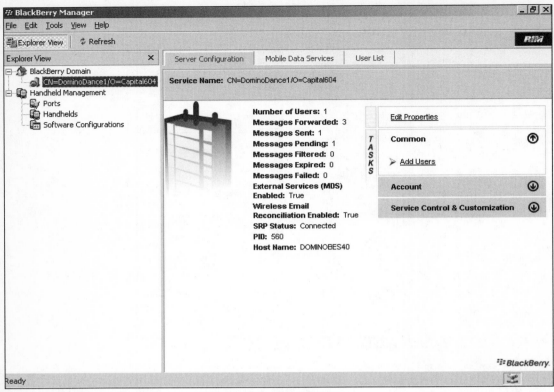

Figure 6-2: Selecting the server name before configuring BESAlert

On the next screen, click BESAlert, as shown in Figure 6-3. Type the required information for BESAlert to send e-mails based on a particular alert level.

On the BESAlert configuration screen, you must enter the following information:

❑ **SMTP Hostname:** DNS name of your SMTP server.

❑ **SMTP Account name:** Used only if your SMTP server requires authentication.

❑ **SMTP Address:** The SMTP address of the person who will receive the BESAlert e-mails.

❑ **Event Level:** The level of event log entries that will be sent.

❑ **User Notification:** Use if you would like to send the same notifications to a management console or another e-mail address.

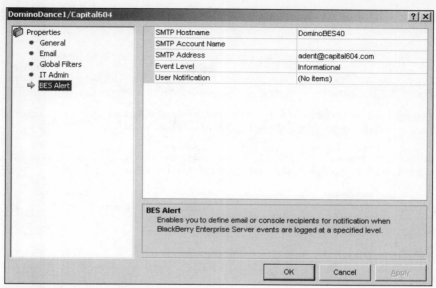

Figure 6-3: Configuring BESAlert options

Using Simple Network Management Protocol

The use of the Simple Network Management Protocol (SNMP) as a method of monitoring your BlackBerry environment represents a step up in functionality from the BESAlert service.

SNMP has two components. The first is the server component that consists of an *agent* that can monitor the server and make configuration changes to it, if necessary. If the agent detects a problem, it sends an SNMP trap event to a predetermined IP address. The second component of SNMP is the SNMP *trap server*. This is a server that runs an SNMP agent and listens for SNMP trap events. It then logs them or acts upon them. The agent would send an e-mail, send a page, or make a network change depending on how it is configured.

To make use of SNMP, you will need a networking tool that can load an SNMP MIB and enable you to choose what indicators you want to monitor. Typical tools used are Observer by Network Instruments and HP Openview. Because SNMP MIB files often provide a variety of indicators, the network tool that you choose will be able to alert you when problems occur, as well as to monitor certain statistical indicators and log them to a file. The log files can be used to monitor the ongoing performance of your BES servers and mail servers.

For an effective, overall perspective of your BlackBerry and mail infrastructures, you will need to make use of the following:

❑   **The BES SNMP MIB:** Has the ability to monitor BES-specific counters.

❑   **The SNMP MIB** (for your Windows servers that host the BlackBerry components and mail servers): A Windows-specific MIB that monitors Windows servers.

❑   **The mail server MIBs** (for Microsoft Exchange, Lotus Domino, and Novell GroupWise): A mail-system specific MIB that monitors your mail servers.

The BES MIB provided on the installation CD (or found in the directory c:\windows\system32) provides 121 counters and traps in the categories shown in the following table.

Counter Type	Number of Counters
Global Statistic	17
Configuration	18
System Health	36
Mail Server Health	6
User Health	37
Trap Type	**Number of Traps**
BlackBerry Enterprise Server Events	7

When using an MIB browser, you must be aware of the Object Identifiers (OIDs). These are numbers that identify each MIB object in the MIB file. You will be able to identify the BlackBerry OID as follows:

❑   BlackBerry Counters begin with 1.3.6.1.4.1.3530.5.

❑   Traps begin with 1.3.6.1.4.1.3530.5.9.

To make it easier to find the BlackBerry counters and traps, you should compile the MIB; otherwise you will have to search for them using their OIDs. If you do not have a program that will compile MIBs, you can start by downloading an evaluation copy of Visual MIB Browser Pro by NuDesign, shown in Figure 6-4. Figure 6-5 shows an example of trending counter information using one of the BES MIB OIDs.

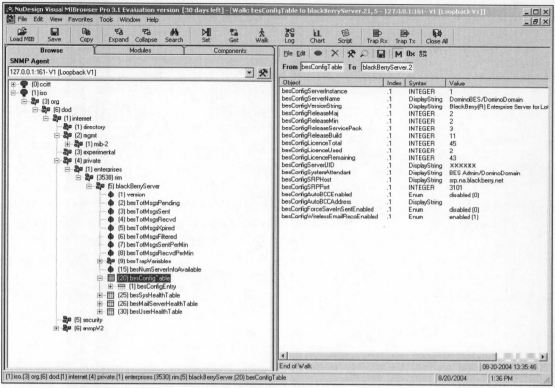

Figure 6-4: Browsing the BES MIB using Visual MIB Browser Pro

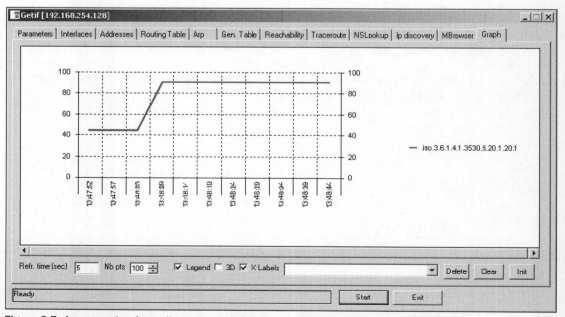

Figure 6-5: An example of trending counter information using one of the BES MIB OIDs

PerfMon

If you do not feel comfortable working with SNMP, you can make use of a proprietary Windows tool called *PerfMon*. This tool provides the capability to monitor hundreds of indicators on a Windows server. Each time you install more software on that server, that software's indicators are added to the list.

After your BES has been installed you will be able to gain access to the new PerMon indicators shown in the following table.

Indicator	Description
Connection State	Shows whether the BES has an active connection to the wireless network.
Messages Expired	Indicates the total number of expired messages for this session.
Messages Filtered	Indicates the total number of messages to which filters have been applied and not been forwarded to the handheld devices.
Messages Queued for Delivery	Indicates the total number of pending messages awaiting delivery to the handheld devices.
Messages Received	Indicates the total number of messages received by handheld devices in this session.
Messages Received/min	Indicates the average number of messages received per minute by handheld devices in this session.
Messages Sent	Indicates the total number of messages sent by handheld devices in this session.
Messages Sent/min	Indicates the average number of messages received per minute by the handheld devices in this session.

Start the PerfMon by selecting it in the Start menu of your BES server. To gain access to the new BlackBerry-specific performance indicators, click the Add button and choose BlackBerry Server, as shown in Figure 6-6. After it has been selected, you will see the performance indicators listed in the previous table. You can add the ones you want, to monitor to the main PerfMon screen. Figure 6-7 shows an example of real-time statistics using BlackBerry-specific counters in PerfMon.

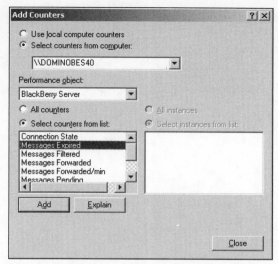

Figure 6-6: Adding BlackBerry-specific counters to PerfMon

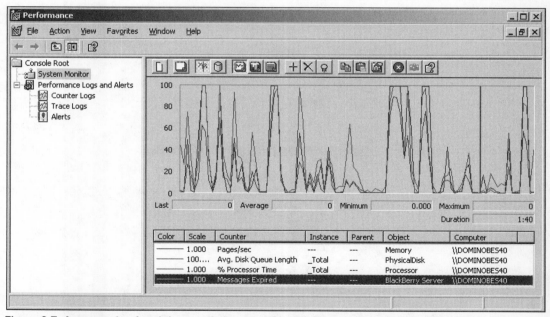

Figure 6-7: An example of real-time statistics using BlackBerry-specific counters in PerfMon

While it is cool to watch these performance indicators in real time, it is much better to set up a trace so that the indicators are logged to a file for later review and for trend analysis. To do this, right-click Counter Logs and select New Log Settings. When you are setting up your new Counter Log trace, you will be able to add objects or counters (see Figure 6-8). If you add an Object and select the new BlackBerry Server object, all counters relating to the BES will be collected. If you select Add Counters, you will be able to manually select the counters of your choice (see Figure 6-9).

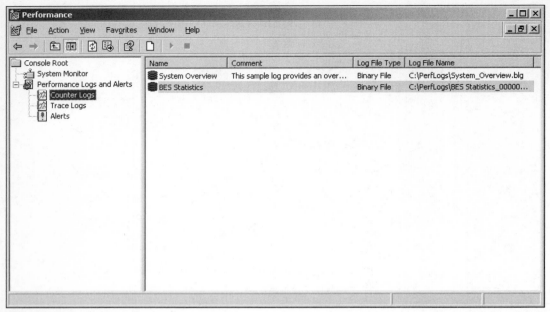

Figure 6-8: Creating a new trace file

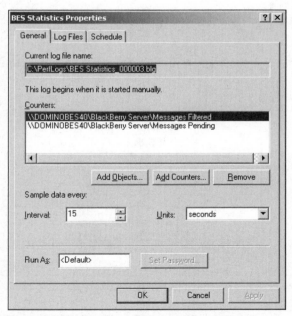

Figure 6-9: Adding BlackBerry-specific counters to the trace file

By using PerfMon and by gathering logs through SNMP, you can create a baseline for server performance that enables you to keep track of the changes in that performance over time. (See the section "Load Balancing BlackBerry and Mail Servers" later in this chapter for more information on base lining.)

Monitoring Mobile Data Service

You can monitor the performance of a Mobile Data Service (MDS) by connecting to its administration interface. To connect to the administration interface, browse to the following URL.

```
http://<MDS_computer>:<web_server_port>/admin/common
```

Following are the variables included in this URL:

❑   <MDS_computer>: This represents the DNS name of the server where you have MDS running.

❑   <web_server_port>: This represents the MDS port number (typically 8080).

Here is an example:

```
http://NYMDS:8080/admin/common
```

After the screen is loaded, you should see something similar to Figure 6-10.

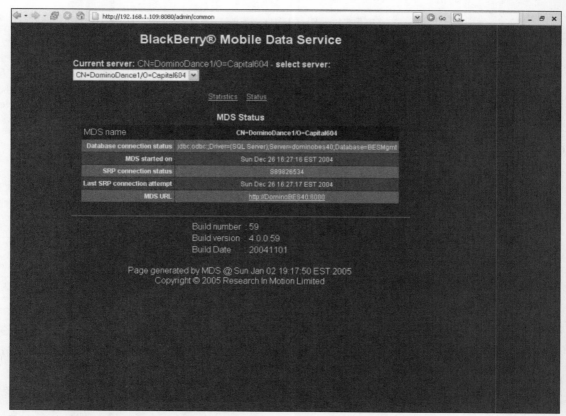

Figure 6-10: Administration interface for MDS

It is good to take a look at the performance of your MDS from time to time to ensure that it is not over-utilizing the server. If this occurs, you should consider moving it to a separate server.

Using System Logger

System Logger (SysLog) is an industry-standard protocol for capturing events from devices on a network. Devices that experience an event simply send the information about that event in a standard format to a predetermined UDP port on a predetermined host.

When the SysLog host receives a SysLog event, it will take some action. Most of the time, it will log the event to a file. Depending on the program you use, it can send you an e-mail, a page, an instant message, and so on. You can use this system to keep track of events on your hosts (including your BES servers), or to alert you when a certain event occurs.

To make use of SysLog, first install a program that listens for SysLog events on a server other than your BES servers. Programs that listen to SysLog events are Hp Openview, Argent Guardian, and so on. Configure the BES to send SysLog messages to this server. By default, the BES will send these SysLog alerts to UDP port 514, so you can either set up the SysLog software to listen on that port, or change the BES configuration accordingly.

To configure the BES for SysLogging, use the `regedit` utility on the BES server to edit the registry on your BES server(s). Browse to the following registry entry:

```
HKLM\Software\Research In Motion\BlackBerry Enterprise Server\Logging Info\
```

Before making any changes to your Windows registry, be sure to back it up first.

Under this key you will see all of the components that you can set to send SysLogs. Under each key you must modify two values. First, modify the SysLogHost value (a string value) and type in the IP address of the SysLog host and the port number (if you want to use something other that port 514). For example if you wanted to have this component send its SysLog data to `192.168.1.10` on port 604, you would enter the following:

```
192.168.1.10:604
```

Next, modify the `EventLogLevel` (a dword value) value and enter the log level you would like to use for your SysLog events. The log levels are as follows:

- ❏ 1: Error
- ❏ 2: Warning
- ❏ 3: Informational
- ❏ 4: Debug
- ❏ 5: Other (5s are always logged)

By default, all of the BES components are configured with `EventLogLevel` equal to 2 (Warning).

Repeat this process for each component key.

Load Balancing BlackBerry and Mail Servers

As previously mentioned, adding BlackBerry devices to your e-mail environment has an impact on the mail servers themselves. In addition, the BES servers can become overwhelmed. This section discusses which statistical indicators you must baseline and track to ensure that you keep your BlackBerry and mail servers performing at their peak.

Baselining Your Servers

Now that you know the various options you have to monitor your BES, you should use one or more of these tools and methods to baseline your BES server performance. Three factors must be considered when ensuring good performance on any given Windows server:

❑ **Processor capacity:** A Windows server is performing well when the processor usage does not exceed 75 percent. This is true when using multiple-processor servers. Each processor's usage must not exceed 75 percent.

❑ **Disk input/output bandwidth:** The disk input/output bandwidth is measured by the *disk queue length*. This measures the number of read or write requests (made to the disk by the operating system) that have not yet been completed, and, hence, are sitting in the disk queue. A typical Windows server today is connected to a disk array that is made up of multiple disks. Knowing how many disks are in your server's array will help you calculate whether your server's disk queue length is good or bad. A disk queue length greater than two per disk (or spindle) indicates that your disk array has a bottleneck if this server is a Lotus Domino mail server. A queue length greater than one per disk (or spindle) indicates a bottleneck if the server is a Microsoft Exchange server.

❑ **Available memory:** Memory usage is also an important factor in server performance. Lotus Domino manages its own memory, so watching indicators such as the Memory Pages Per Second will not offer an insight into a possible memory bottleneck. Instead, you should ensure that your server does not approach zero available physical memory. Microsoft Exchange will use up all available memory, so watch the Memory Pages Per Second indicator. The pages per second must not go higher than 10, and should stay at zero or return to zero quickly.

Begin your baseline by monitoring these performance indicators on every mail server before you add any BlackBerry users. If you already have BlackBerry users, use these indicators to ensure that the configurations of your current mail servers are able to handle the load.

Save the performance data for future use as a reference and keep monitoring these indicators daily. As you add BlackBerry users, look to see how these performance indicators increase. You should see a trend developing that will enable you to predict when you must make adjustments to your environment to compensate for the increased demand on your mail servers.

Based on your trend analysis, you may choose to add an additional mail server and start moving users off to the new server. Keep in mind that you should baseline this new server and watch its indicators moving forward.

Monitoring Your Servers

Your BlackBerry server must be monitored. As you add more and more users to it, based on how those users work with their BlackBerry devices, the load will increase at different rates. Monitor the same performance indicators as you do for your mail servers, but take into account that the server on which you have installed the BES might also have the Attachment Service, MDS, BlackBerry Router, and BlackBerry Dispatcher installed.

If you notice that your processor usage is approaching 75 percent per processor, you should monitor how much processor time is being taken up by the different components. If your BlackBerry users often view attachments on their handheld devices (especially large attachments), the Attachment Service will start using up more processor cycles. It may be advantageous to move the Attachment Service to a separate Windows server.

The same goes for all of the other BlackBerry components. Plan to move them to different servers if they seem to be using up more processor cycles. If you are using a BES running on Lotus Domino, the BES is an add-in Domino task, and so the processor usage will show as coming from NSERVER.EXE (the main Domino service). If you are using a BES running in a Microsoft Exchange environment, you will see that the processor usage is coming from the BES task.

If you see that the process using up the majority of the processor cycles is the BES task, you should consider installing a new BlackBerry server and moving users off the old one onto the new one. This will have the effect of easing the load on the BES servers.

Summary

This chapter discussed the ways in which you can monitor your BlackBerry environment. You have learned how to receive e-mail or pager alerts when certain conditions are met, so that you can take the appropriate actions to rectify them. The monitoring methods include the use of BESMonitor, BESAlert, SNMP, PerfMon, MDS, and SysLog.

In addition to learning how to monitor your BlackBerry environments, you learned the importance of performing a baseline of certain statistics. These baselines help you determine whether your BES and mail servers are performing within the optimal parameters, and whether you must further balance the load by rearranging users or services.

In Chapter 7, we discuss the different ways of managing your BlackBerry users. We will discuss this topic in the context of BES 4.0 and older versions of BES.

Managing Your BlackBerry Users

As a BlackBerry administrator, your responsibilities include controlling your BlackBerry environment. You can control what users can do on their devices, who can push content out to those devices, and even help your users when they lose or break their devices.

You have many means at your disposal to accomplish these goals. This chapter discusses several of the tools you can use to help manage the users in your BlackBerry environment, including the following:

❑ **IT commands:** Commands that are sent to the device wirelessly, that instruct it to perform a certain function.

❑ **PIM settings:** Control how the users' devices synchronize with their PIM data.

❑ **Wireless synchronization:** Enable or disable the ability for the users' devices to synchronize email and PIM data wirelessly.

❑ **Redirector settings:** Modify the user's signatures and e-mail filters, if necessary.

❑ **MDS access control:** Control how the user can use MDS.

Throughout this chapter, we will use screenshots from the Lotus Domino BES BlackBerry Manager. If you are using an Exchange BES 4.0 BlackBerry Manager, the GUI is slightly different. However, the settings and their locations are the same.

IT Commands

By sending IT commands to the devices, you can make the devices perform a certain function. These IT commands were first introduced in the Domino BES 4.0 product, so they are not available on the Lotus Domino BES 2.2. For Exchange BES environments, the IT commands have been available since BES 3.6. In addition to the correct BES version, the device must be running at least version 3.6 of the Handheld Software.

To use this feature, begin by opening the BlackBerry Manager console. On the far left of the screen, select the desired server. At the top of the screen, select the User List tab, then click on one of the users below the tab. As shown in Figure 7-1, you can view the IT Admin tasks by clicking that heading in the lower-right portion of the screen.

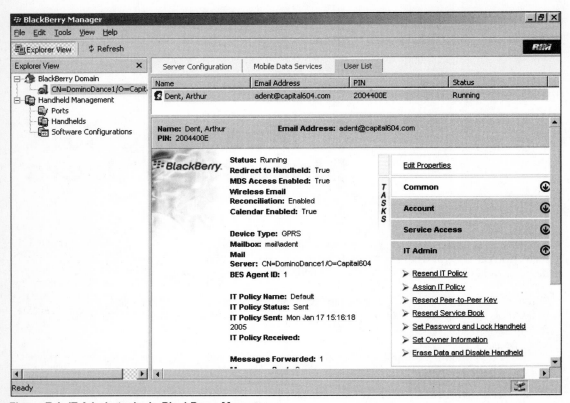

Figure 7-1: IT Admin tasks in BlackBerry Manager

The list of available tasks includes the following:

- ❑ Resend IT Policy
- ❑ Assign IT Policy
- ❑ Resend Peer-to-Peer Key
- ❑ Resend Service Book
- ❑ Set Password and Lock Handset
- ❑ Set Owner Information
- ❑ Erase Data and Disable Handheld

Resend IT Policy

As discussed in Chapter 3, IT Policies are a set of parameters you can use to enforce company policy, security, or simply limit the available functions on the BlackBerry handheld device. In BES 4.0, IT Policies can also contain settings for third-party applications that may be running on your devices.

There should be no need to resend IT Policies because they are automatically sent out when they are updated, or when you assign a new one to a user or users. However, if for some reason a particular device does not receive an IT Policy, you can use this task to resend it.

Assign IT Policy

When a BlackBerry user is added to the BES, the default IT Policy is automatically sent to that user. If a user is moved between BlackBerry Enterprise Servers in the same BlackBerry domain, that user will keep using the same IT Policy. To be safe, the BES will resend the IT Policy to the device. If at some point you create a new IT Policy, you can use this task to assign it to one or many BlackBerry users.

Resend Peer-to-Peer Key

A *peer-to-peer encryption key* is a key that you can set to limit the sending of PIN-to-PIN messages. If you set a peer-to-peer key, then your BlackBerry users will not be able to send PIN-to-PIN messages to devices outside of their BlackBerry domain, although they will be able to receive incoming PIN messages from any device. When you set the encryption key, all PIN-to-PIN communications are encrypted with that key. This is in addition to the 3DES encryption that is applied to all BlackBerry communications. To set a peer-to-peer encryption key, click your BlackBerry domain in the left of the BlackBerry Manager screen, and then click Update Peer-To-Peer Encryption Key under the Service Control and Customization section on the right.

To resend this encryption key, use the Resend Peer-To-Peer Key command in the IT Admin section of the BlackBerry Manager console screen.

Resend Service Book

Service Books are entities that describe to the handheld how to connect to certain services. For example, the MDS Service Book (which is called IPPP) enables the handheld for MDS use and tells it how to access the service. Your cellular carrier will send Service Books to your handheld, enabling the WAP browser, and any Web mail services that you sign up for.

Service Books are always sent from the BES automatically. If at any time you need to resend the Service Books because you believe that a handheld is missing them, you can use this IT Command to do so.

Set Password and Lock Handheld

This IT command sends to the user's device a wireless command that sets or changes the handheld password. In addition, it locks the device, which then requires that the new password be entered to unlock the unit. When sending this IT command, an option enables you to change the Owner Information screen that displays when the device is locked (see Figure 7-2).

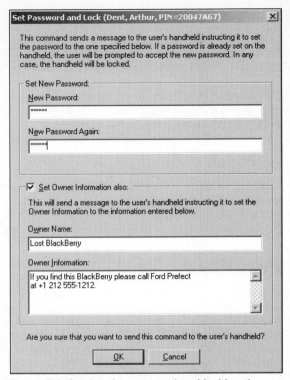

Figure 7-2: Setting the password and locking the handheld unit

This IT command is useful when a user loses a device. As you can see in Figure 7-2, it not only allows you to secure the device, but also to display a message on the device that informs the potential Good Samaritan how to contact you to return it.

Set Owner Information

This IT command sends a wireless command to the device to change the Owner Information screen. As shown in Figure 7-3, the result of this command is very similar to the lower half of the screen shown when sending the Set Password and Lock Handheld IT command.

This command is useful when a user loses a device, but the user is positive that the device is locked with a password. You can change the Owner Information screen to instruct the person who finds the device how to contact you to have it returned.

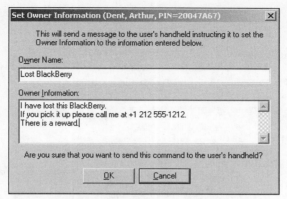

Figure 7-3: Setting owner information

Erase Data and Disable Handheld

This IT command sends a wireless command to the device that causes it to erase all of the user's data and disable it so that it can no longer be used on your BES. Before the data is erased, you will see an alert dialog box, indicating that all data will be erased if you proceed (see Figure 7-4).

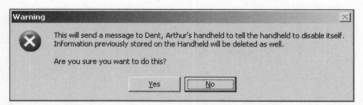

Figure 7-4: Warning indicating that all data will be erased from the handheld

This IT Command is useful in situations when a device has been lost or stolen, and there is no chance of its being returned.

PIM Settings

The Personal Information Manager (PIM) settings contain personal user information that is synchronized with the device. If you are still in a pre-4.0 BlackBerry environment, the PIM settings include the synchronization (through the Desktop Manager to the device) of the Address Book, Memo, and Tasks. In a 4.0 BlackBerry environment, the PIM settings include the wireless synchronization of the e-mail filters, e-mail settings, Tasks, Memo, and Address Book.

With BES 4.0, you can use global settings to set up a global standard by which all users are initially set up, and you can modify user-specific PIM settings to make changes for individual users.

The word global in the BES 4.0 context means that a setting or policy is used for all BES servers that connect to a single SQL database.

To configure the global PIM settings, open the BlackBerry Manager and click the BlackBerry Domain option in the upper left of the screen. On the right of the screen, click Edit Properties, as shown in Figure 7-5. On the next screen shown in Figure 7-6, click Global PIM Sync.

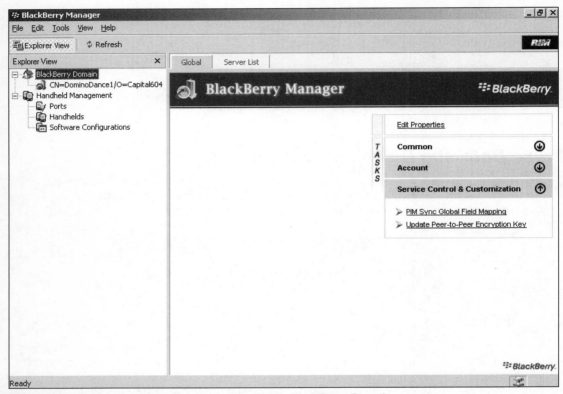

Figure 7-5: Selecting the Edit Properties option for the BlackBerry Domain

You will notice that each available PIM setting is represented in a section of the pane on the right-hand side of the screen. Each PIM setting has the following fields:

❑ **Synchronization enabled:** This enables (set to `True`) or disables (set to `False`) wireless synchronization.

❑ **Synchronization type:** This option sets the synchronization to Bidirectional, Handheld to server, or Server to handheld.

E-mail filters and e-mail settings are the exception to this rule, because they have a bidirectional setting only.

❑ **Conflict resolution:** This controls who wins if a conflict arises between the device and the server. The choices are the handheld unit wins or the server wins. An example of a conflict would be if you moved an e-mail to a particular folder in your e-mail client and also moved that same e-mail to a different folder on your device before it wirelessly synchronized. If you set the conflict resolution to Server Wins, then the folder move that you performed on the device will be overridden; the folder move that you performed in your e-mail client will take precedence.

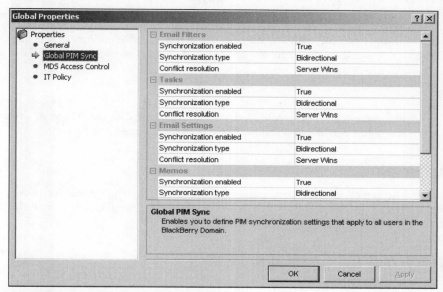

Figure 7-6: The Global Properties screen

E-mail Filters

E-mail filters are the settings that enable you (or the user) to control which e-mails are actually sent to the device. Some users choose to use the e-mail client's rules to move e-mails directly to a particular folder based on a subject line or sender. However, other users would rather have those e-mails sent to their Inboxes and set a BlackBerry filter so that the e-mails are not copied to their devices. With BlackBerry 4.0, the user has the ability to make changes to his or her BlackBerry filters right from their device. If you want to allow users to be able to make these changes from their devices, you must enable wireless synchronization for the e-mail filters.

Tasks

This PIM setting controls whether the user is able to synchronize tasks (sometimes called *ToDos*) wirelessly between an e-mail server and a device.

E-mail Settings

E-mail settings affect the control the user has over mailbox and BlackBerry settings. As an administrator, you establish this control through the PIM settings.

Chapter 7

In a BlackBerry 4.0 environment, the device has a new menu called Email Settings (located under Messages, Options). Under this menu, the user can choose to have e-mail delivered to a device. The user can choose to disable the e-mail if he or she goes on vacation and does not want to receive company e-mail, but still wants to receive personal e-mail or continue to use the phone and Web browser. The user can also choose whether to save a copy of all e-mail sent from a device into his or her mailbox Sent folder.

Control over whether to use an automatic signature and the ability to completely edit that signature are offered under this new menu. Finally, the user can set an out-of-office notification message, as well as indicate when it will expire. As an administrator, you can allow your user to modify these settings.

Memos

Memos on the BlackBerry can synchronize with the Notes Journal, Outlook Notes, or GroupWise Posted Notes. This setting controls whether it can be wirelessly synchronized and how.

Address Book

The BlackBerry Address Book can synchronize with the e-mail client's Personal Address Book. This setting controls whether the Address Book can be wirelessly synchronized and how that synchronization is configured.

User-Specific PIM Settings

While you can modify the Global PIM Settings that apply to all new BlackBerry users being added to any BES connected to the same SQL database, you can also modify these same PIM settings on a user-by-user basis. In addition, you can enable or disable wireless synchronization altogether. In addition to the PIM settings, you can enable or disable *wireless backup*. Wireless backup is a feature that wirelessly backs up the user's device preferences such as ribbon icon positions (which are the position of the icons on the BlackBerry Home Screen), browser bookmarks, audio profiles, and so on.

Setting Up Full Wireless Synchronization

In the BES 4.0 environment, your users have real-time access to all of their PIM data. They may also have that data updated on their devices, as well as on the server.

To achieve this in the Exchange BES environment, since the PIM data is already stored on the Exchange server, you merely enable the wireless PIM synchronization for each PIM setting. In a Domino BES world, it is slightly more complicated to make the data available on the server so that the Synchronization Service can update it. The three methods available to you include the following:

❏ Domino Roaming User profiles

❏ iNotes

❏ Manual replicas

Using Domino Roaming User Profiles

From Domino 6.0 and beyond, a new feature called *Roaming Users* provides the capability to automatically create replicas of each user's Personal Address Book (names.nsf), the Journal (journal.nsf), and the Notes bookmarks. These files are all stored on the local computer, and the Roaming Users profile allows replicas of these databases to reside on a Domino server.

If you enable Roaming Users for a particular Notes user, a sequence of events takes place that causes the user's Notes client to create the replicas of the databases on a designated Domino server. This feature also sets up a replication schedule so that the databases are always kept in synch. If you enable Roaming Users for a Notes user, and then enable Roaming Users for wireless synchronizing on the BES, the BES will automatically use the information in the Roaming Users section of that user's Person Document (each Notes user's configuration document that is stored in the Notes Directory) to determine where to find the Memo Pad (journal.nsf) and Address Book (names.nsf). If you enable Roaming Users for a particular user, and then enable Roaming Users to synchronize their Memo Pad and Address Book wirelessly, you have no control over where the BES looks for these databases.

Using iNotes

iNotes (a new feature added in Domino R5) is an enhanced Web interface to a user's e-mail and calendar. Each user has the capability to synchronize an Address Book and Journal with iNotes. In the Notes client, the user goes to Actions ➪ iNotes Web Access and synchronizes the Contacts and Journal. This puts the data into a special location in the user's mail file so that it can be accessed when the user uses iNotes.

Because that data is now in the user's mail file, BES 4.0 can get to it. The user must perform the synchronization steps regularly so that the data stays current.

Using Manual Replicas

If it is not possible for you to use either iNotes or Roaming Users to have server replicas of the user's Address Book and Journal, then you can manually create the replicas. You could send a team of technicians around to each user's desk to create the manual replicas, or instruct the users on how to do this. Either way, you must adjust the security setting for the user so that the user can create server replicas.

You will need to create some kind of directory-structure model or file-naming model to deal with these replicas. For example, you could create a new directory on a designated Domino server called PIM, and under that a directory for each user by first initial last name (for example, e:\lotus\domino\data\pim\adent, which would be Arthur Dent's directory). In each directory, you would have two files, journal.nsf and names.nsf.

Another approach would be to use one directory, and rename the database replicas using the user's name. In this instance, you would create a new directory called PIM as in the previous example, but no subdirectories. Each user would then replicate the names.nsf and journal.nsf files to this directory, but rename the files based on the user's name (for example, e:\lotus\domino\data\pim\adent_names.nsf and e:\lotus\domino\data\pim\adent_journal.nsf). You could use variations of this such as adent_ab.nsf for Address Book and adent_j.nsf for Journal.

When the files are accessible, manually set up wireless synchronization using the BlackBerry Manager. Edit the properties for each BES user and enter the appropriate server and file location information. The server information is in the following format:

```
CN=<servername>/OU=<servers>/O=<companyname>
```

The file location is entered using the path to the database without any preceding back slashes, as shown here (see Figure 7-7):

```
PIM\adent_journal.nsf
```

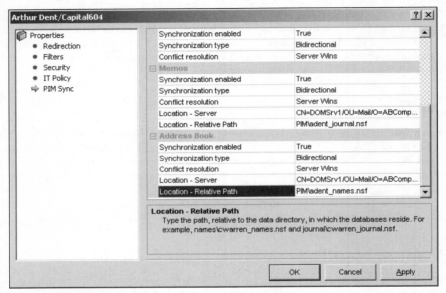

Figure 7-7: Manually setting up wireless synchronization

One important tip when using wireless synchronization is that you cannot mix methods. For example, if you have set up a user as a Domino Roaming User, you cannot enter the location of the Address Book and Journal manually. This is because every 15 minutes the BES will refresh the data from the company Address Book, and the locations of the database will be overwritten with that data. If one of your users has used iNotes to view the Address Book or Journal, then the BES will default to using the data found in the mail file. If you then make the user a Roaming User, the BES will start using the data in the Roaming User profile. However, if a user has used iNotes in the past, and you try to specify a manual location to the Address Book or Journal, while it will look correct on the BlackBerry Manager console, the Synchronization Service will become confused and the initial synchronization of data will fail.

It is best to plan your wireless synchronization setup ahead of time and clean up any iNotes data before attempting to use a manual method.

Redirector Settings

Redirector settings used to be set through the Desktop Manager. However, in BES 4.0, you are now able to control them from the BlackBerry Manager. If you allow it, each user can control the redirector settings from the BlackBerry device, but there may be a reason why you will want to make changes to the redirector settings on behalf of the user. These settings are also useful if the user's device is still running a version prior to BES 4.0 and the user needs these settings changed.

To access redirector settings, double-click the user name in the BlackBerry Manager.

As you can see in Figure 7-8, the following options are available:

❑ Redirect to Handheld

❑ Do Not Redirect When in Cradle

❑ Do Not Save Sent Messages

❑ Signature

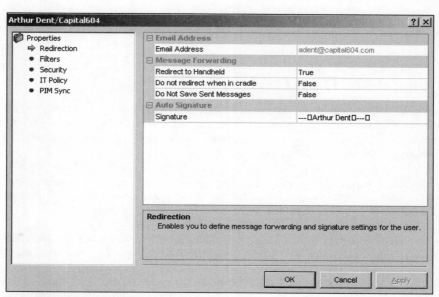

Figure 7-8: Redirector settings

Redirect to Handheld

This setting controls whether e-mail from the user's mail file is sent to the device. The option is either True or False. Typically, this option is changed when the user is out of the office and does not want to receive e-mails on the BlackBerry wireless device but still wants to continue using the phone or Web browser. If, for some reason, the user cannot change this option on the device, you have the capability to control it using the BlackBerry Manager console.

Do Not Redirect When in Cradle

This option can only be enforced if the user is running the Desktop Manager. If so, you can set this option to govern whether e-mail is sent to the device when it is connected to a computer. It is possible that the user does not want a mirror image of the Inbox on the device and, while the user is at his or her desk, the user does not want e-mail that has already been responded to on the device. If you do require a Desktop Manager for all users (because they need to synchronize with PIM data that is not from the e-mail client), but your users want to receive all e-mails on their devices no matter whether it is connected to their computers, then you would set this option to False.

Do Not Save Sent Messages

By default, this option is set to False, which means that all e-mail sent from the user's device is saved in the mailbox Sent folder. Sometimes, because of security or retention policy requirements, you may want to set this option to True so that e-mail is not saved in the Sent folder.

Signature

This is the signature that is appended to the user's e-mails by the BES. You may want to modify this signature if the one that the user has chosen to use is inappropriate.

MDS Access Control

In pre-4.0 BlackBerry environments, the use of MDS was not controlled. Any BlackBerry user who had MDS enabled for an account could access any Web site on the company intranet, as well as the Internet. Anyone could also push out (that is, send) content to any device by just knowing how to write a few lines of code. (Chapter 9 discusses MDS in more detail.)

In the 4.0 BlackBerry environment, you have much tighter control over this *push* (sending) and *pull* (retrieving) of data. To control who can push and who can pull data through MDS, as well what kind of data can be pulled, is controlled by roles, authentication, and URL patterns.

We will first look at the configuration screen and examine each option. Click the BlackBerry Domain and then click Edit Properties. On the next screen click MDS Access Control, as shown in Figure 7-9.

Notice that the right-hand side of this screen is broken up into two parts: Pull and Push.

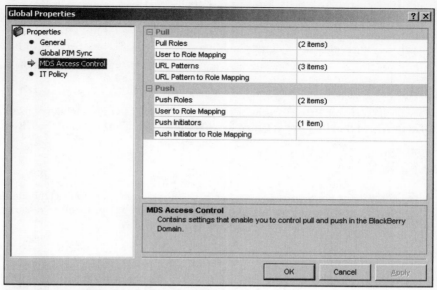

Figure 7-9: The MDS Access Control option

Pull

As mentioned previously, *pull* refers to the capability of a BlackBerry user to pull content from the Internet or company intranet through MDS to a device. This section of the MDS Access Control screen provides the following configurable fields:

❑ Pull Roles

❑ User to Role Mapping

❑ URL Patterns

❑ URL Pattern to Role Mapping

Pull Roles

Using this option, you can create, remove, or modify Pull Roles. Technically speaking, a *role* is just a name and description. It has no functionality on its own. Note that the field contains the phrase *(2 items)*. This refers to how many of each item has been previously created. If you click the field, an ellipses icon appears. When you click that icon, the items will be listed, allowing you to choose them.

User to Role Mapping

After you have created some roles, you can use this option to map your BlackBerry users to the roles. This is similar to adding members to a group in Microsoft Active Directory Services or Novell NetWare eDirectory.

URL Patterns

Here you can define certain URL *patterns*, which signify what URLs your BlackBerry users can browse to through MDS provided that they are assigned to a particular role. The format of the URL pattern is as follows:

```
<hostname:port/path>
```

As you can see, three items are required in this field. The hostname refers to the DNS name of the host server, port is the TCP port used, and the path is a path to the Web page or other resource. Following is an example:

```
portal.abcompany:80/index.html
```

You can use an asterisk (*) as a wildcard. This indicates that all pages are accessible on that host. Following is an example:

```
portal.abcompany:80/*
```

When using URL patterns, you can choose the type of service. Those service types are http, https, ldap, tcp, and ocsp. (Note that OCSP stands for Online Certificate Status Protocol, which is a protocol designed to provide real-time certificate validation.)

URL Pattern to Role Mapping

This is where you put it all together by creating a mapping (or link) between the URL pattern and the roles. For example, if you wanted a BlackBerry user called Arthur Dent to have access only to www.google.com, you would follow these steps:

1.   Create a new role called Google Access.
2.   Assign Arthur Dent to that role.
3.   Define a URL pattern (google.com:80/*).
4.   Map the URL pattern to the role called Google Access.

Push

As mentioned previously, *push* refers to the capability of a BlackBerry user to receive content pushed to their device. This content includes a channel, Web message, and browser cache data (which is discussed in greater detail in Chapters 11 and 12). Push also refers to the capability of someone on your company intranet to push this type of content to the devices.

This section of the MDS Access Control screen, shown in Figure 7-9, provides the following configurable fields:

❑ Push Roles

❑ User to Role Mapping

❑ Push Initiators

❑ Push Initiator to Role Mapping

Push Roles

Like Pull Roles, Push Roles are just a name and description. They have no functionality on their own, but are equivalent to groups. Again, note that the field contains the phrase *(2 items)*. This refers to how many of each item has been previously created. If you click the field, an ellipses icon appears. When you click that icon, the items will be listed, allowing you to choose them.

User to Role Mapping

This is where you map a BlackBerry user to a role.

Push Initiators

Someone who sends data to the device through a push method is called a *push initiator*. In this field, you can define push initiators and the password that the push initiator must use to perform a certain kind of push request. This helps limit who can actually send data to your devices.

Push Initiator to Role Mapping

This is where you tie the role to the push initiator. This effectively tells the BES that the BlackBerry users in a particular role can accept push data from a push initiator, as long as that initiator uses a particular password.

Enabling Push and Pull Control

After you have set up push or pull control, you must configure the BES so that it honors the settings. To do this, begin in BlackBerry Manager. Click the BES server name on the left side of the screen, and then click the Mobile Data Service tab. Finally, click the Edit Properties button.

On left side of the next screen (see Figure 7-10), click Local Access Control. On the right side, you will see Pull Authorization. If set to `True`, the BES will start enforcing the roles and URL patterns. You will also see Push Authentication. If set to `True`, the BES will start limiting the push requests to the push initiators.

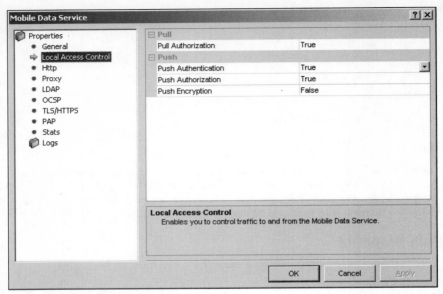

Figure 7-10: Local Access Control for MDS

If you also set the Push Authorization to True, the BES will start enforcing the Push Initiators and Push Initiator to Role Mapping settings, which will limit the push data to certain push initiators going to certain groups of users.

The Push Encryption field shown on this screen specifies whether the push requests are encrypted in any way. Once set to True, the push requests will be encrypted using Transport Layer Security (TLS) or Secure Socket Layer (SSL).

Summary

Managing BlackBerry users became much easier with the introduction of BES 4.0. This chapter has explored the different ways in which you can manage your BlackBerry users and manage who can push content to them.

Even if your users' devices have not yet been upgraded to version 4.0, you can still manage your users from the BlackBerry Manager by changing filters, signatures, whether the users receive e-mail, and if devices receive email while in the cradle.

You can also now control who can push content out to your user's devices, and who can receive that content, which is a great step forward in securing your devices.

In Chapter 8, we discuss methods of disaster recovery. We will examine how to set up your BlackBerry environment in such a way that it can be made available in the event of a disaster or other downtime.

Disaster-Recovery Planning

This chapter addresses what should be a primary concern: how to plan for a disaster or an event that disables your BES. With proper planning, you can ensure that your BlackBerry users do not lose any functionality during BES downtime. We will discuss multiple methods of disaster recovery. As you will see, how quickly you want to return full functionality to your users after a disaster or BES outage will affect how much you spend on the solution and how complex it is.

The discussion in this chapter focuses on the following two BES environments (at the time of writing, RIM has not yet released any proposed disaster-recovery scenarios for Novell GroupWise):

❑   Lotus Domino

❑   Microsoft Exchange

Lotus Domino BES

As discussed in previous chapters, the Domino BES implementation relies on State Databases to keep track of data that has been sent to the handheld device. These State Databases become critical when planning for a disaster or outage, because without them, the data on the handheld unit becomes orphaned.

We will discuss disaster recovery for Domino BES in two parts. The first part will cover disaster-recovery scenarios for pre-4.0 BES implementations, and the second part will cover disaster-recovery scenarios for BES 4.0.

Pre-4.0 Domino BES (BES 2.2)

With Domino BES versions prior to BES 4.0, there is no SQL database that stores the user configurations. Instead, there is a BlackBerry Profiles database (`blackberryprofiles.nsf`) on each BES. There is also a database that tracks MDS statistics (`bbstats.nsf`), an outgoing queue database (`blackberryoutbox.nsf`), a Directory Database that is used by MDS when sending content to the handheld device (`bbdir.nsf`), and the State Databases (one for each BES user).

Chapter 8

In addition to the databases, there is a unique SRP ID for each BES and an SRP Authentication Key that must stay the same throughout this process. This information is stored in the notes.ini file (found in the \Lotus\Domino directory).

As you will recall from previous chapters, each BES has a unique Server Resource Protocol (SRP) ID number. This number is used to identify the BES to the BlackBerry NOC. Each BlackBerry device communicates through the BlackBerry NOC to its home BES using its SRP ID. SRP is also the communications protocol used to communicate between the BES and the BlackBerry NOC.

In this section, we will examine the following disaster-recovery solutions:

- ❑ Backup and restore
- ❑ Cold standby
- ❑ Knife-edge cutover

Backup and Restore

The simplest way of planning for a BES outage is to back up the data and restore it on new hardware. This is the most time-consuming of the disaster-recovery scenarios because it may require that you build a new server and install the Windows operating system first before any restores can begin. If you keep a standby server with the Operating System (OS) already installed, this may lessen the downtime.

The files that you must back up on a Domino BES (with the exception of the Outgoing Queue) are all of the files in the \Lotus\Domino\Data\BES directory, including the subdirectories. There should only be one subdirectory called \state, which is where all of the State Databases are stored. You must also back up the notes.ini file that contains the SRP ID and SRP Authentication Key.

You can perform a complete system backup and restore your server to the state it was in as of the last backup. You could install Windows and Lotus Domino on a new server and just restore the contents of \Lotus\Domino\data\BES. You could either use the original server ID and notes.ini file to cause the new server to be an exact replica of the original, or you could simply insert the original SRP ID and SRP Authentication Key so that the BES can start up. The BES is not tied to the Windows or Domino server name and therefore you can start it up on any server. For this to function correctly, the rebuilt server must be a mirror image of the original server. It must reside in the same Domino domain and same Notes Named Network as the original.

The drawback of this method of disaster recovery is the fact that the server can be restored only to the state it was in as of the last backup. Any e-mails and other data that was recorded between the backup and the downtime will be lost, which could cause the BlackBerry users to receive duplicate e-mails.

Cold Standby

A *cold standby server* is a server that is kept in a disaster-recovery location and that is running at all times. This server has the same version of Windows installed as the original server, it has the same version of Lotus Domino installed, and it has the same version of the BES installed (using the same SRP ID and SRP Authentication Key). This standby server must reside in the same Domino domain as the original server, and it must also reside in the same Notes Named Network as the mail servers. The cold standby server, however, does not actively run the BES task, and you should remove it from the notes.ini file. The final step is to deny user access to the standby server by editing the server document. Remember that a *server document* in Domino is a document stored in the Domino Directory that specifies how each server is configured. There is one server document for each Domino server in the Domino Domain.

In Domino, you set up one-way manual replication tasks that push new changes from all BES databases (except the Outgoing Queue) to the cold standby server. Each time you add a new user to the primary BES, you must manually replicate the new State Database to the cold standby server. If a user is removed from the primary BES, you must manually remove that user's State Database from the cold standby server. This last step is not absolutely necessary, but is good practice to prevent the Domino server from wasting time indexing unused databases.

In the event of a disaster or downtime, you can start the BES task and all the BlackBerry services on the cold standby server. Depending on the replication interval you set up, the cold standby server's BES should pick up almost at the same point in time where the original BES left off.

When the disaster or downtime is over, you must remember not to allow the primary BES to start up. If it does, there will be two BlackBerry Enterprise Servers using the same SRP ID, and RIM will be forced to disable that SRP ID, causing all BlackBerry users to be without e-mail. The correct way to restore the original BES is to start the Domino server off the network. Quit the BES task and then connect it to the network. Quit the BES task on the cold standby server and perform a one-way replication of all BES databases on the primary server. After that has been completed, start the BES task on the primary server.

Knife-Edge Cutover

The *knife-edge cutover* is the most efficient disaster-recovery scenario and the one that poses the least downtime. This scenario makes use of Domino clustering. While the BES does not support Domino clustering, you can still use it to create the knife-edge cutover.

To set up the knife-edge cutover, you must build a second Domino server with the same version of Domino as the primary. Install the same version of BES using the same SRP ID and SRP Authentication Key, but prevent it from starting automatically. To prevent the BES task from starting automatically, remove it from the `notes.ini` file. Next, you must prevent this second server from accepting any user connections. Change the server document to deny access to all users.

Domino *clustering* is a method of redundancy that provides close to 100 percent uptime. The concept is that you place two or more Domino servers in a cluster. The cluster members will contain identical replicas of certain databases that are kept up to date instantly as changes are made to them. In addition, the clusters support load balancing. Load balancing allows you to configure certain thresholds (such as the number of users connected to one cluster member) and, when that threshold is met, load balancing occurs. In the example of the number of connected users, when the number of users connected to one cluster member has been reached, future users will be redirected to another cluster member to connect to their database.

When setting up a knife-edge cutover, because these servers are in a Domino cluster, you must also disable load balancing on both servers, otherwise the users will be moved to the cluster server when the load balancing threshold is reached.

This is important because if these steps are not taken, the user's Desktop Manager clients will obey the standard Domino clustering rules and connect to the cluster member when their primary server becomes busy. If this happens, you will see replication conflicts in the BES databases, which will cause a number of problems for your users. It is best to avoid replication conflicts in BlackBerry databases at all cost.

Create a Domino cluster and place both the primary and secondary server into the cluster. Make sure that your cluster configuration allows for the automatic replication of changes to the databases. This will cause any changes to be immediately written to the databases on the secondary server.

In the event of a disaster or downtime, the secondary server will be in exact synch with the primary. When you start the BES task and all of the BlackBerry services, the BlackBerry users will not notice any difference. Remember to change the server document of the primary server to deny access to all users, and change the server document on the secondary server to allow all users.

When it is time to switch back to the primary server, you must start that server off the network and quit the BES task. When the BES task is not running, connect it to the network and allow the database replication to sync up. When it is in sync, quit the BES task on the secondary server and allow the replication to sync for a few more minutes. Then start the BES task on the primary server.

BES 4.0

As discussed in previous chapters, the architecture of the Domino BES changed drastically in version 4.0. A relational database is now used for all server and user configurations. That database can either be Microsoft Database Engine (MSDE) or a full SQL server. In addition, there are three new services in BES 4.0:

❏ The BlackBerry Router (which handles least cost routing between the BES and the handheld devices)

❏ The BlackBerry Synchronization Service (which performs all PIM synchronization)

❏ The BlackBerry Policy Service (which manages the IT Policies)

Many of the databases that were found in the Domino 2.2 BES are gone, including the Outbound Queue, the BlackBerry Directory, and the BlackBerry Statistics databases. The BlackBerry Profiles database remains along with State Databases for each user. Finally, the SRP ID and SRP Authentication Key (formerly stored in the notes.ini file) are now stored in the new Configuration Database. With this information in mind, let's examine the following disaster-recovery options:

❏ Backup and restore

❏ Cold standby

❏ Knife-edge cutover

❏ Moving users

Backup and Restore

This option is good for small BES installations where you have chosen to use MSDE as your Configuration Database, and it is installed on the same server as the BES. Technically, you could install and use a full-blown SQL server for your BES Configuration Database.

You should back up the following components:

❏ **The BlackBerry Configuration Database (BESMgmt):** Use a SQL backup job if you are using SQL.

❏ **The BlackBerry Profiles database (**`BlackBerryProfiles.nsf`**):** Most of the contents of this database are synchronized from the SQL or MSDE database. However, the user's encryption keys are stored in this database directly and, therefore, it must be backed up. This database is located in `\Lotus\Domino\data\BES`.

❏ **User State Databases:** These are located in `\Lotus\Domino\data\BES\state`.

❏ **The BES Registry settings:** These are located in `HKLM\Software\Research In Motion`.

❏ **The BES log files:** If you would like to preserve the log information for future troubleshooting, the log files are located in `C:\Program Files\Research In Motion\BlackBerry Enterprise Server\Logs`.

To restore the data to a new server, first build the server and install Windows. Then install MSDE or SQL (if you used SQL). Next, install the same version of Lotus Domino that your original server was running, and use the same server name and ID file. Finally, install the same version of BES that you had running on the original server. Do not be concerned if it creates a new BESMgmt database because you will restore over it later.

Ensure that the BES task is not running and restore the Configuration Database (BESMgmt) to MSDE or SQL. Then, restore the BES files that you originally backed up to `\Lotus\Domino\data\BES`. This restore must include the subdirectory `\state` and all State Databases.

After you have restored the Domino databases, you must run some Domino maintenance utilities on them. Run `fixup`, `updall`, and `compact` on all databases in the BES directory and its subdirectories. After this has completed, start the BES server. When the server restarts, check that all of the related BES services have started. These include:

❏ The BlackBerry Attachment Service

❏ The BlackBerry Controller

❏ The Mobile Data Service

❏ The BlackBerry Policy Service

❏ The BlackBerry Router

❏ The BlackBerry Synchronization Service

Because there is likely to be a number of hours between when the data was backed up and when it was restored, some users may have e-mails on their handheld devices that they can no longer reply to or forward. This occurs because their restored State Database has no record of these e-mails.

Cold Standby Server

The idea of a cold standby server is the same as it is for a BES 2.2. It is a standby server in a disaster-recovery location that is ready to take over from the primary server immediately. However, since BES 4.0 uses MSDE or SQL to store the configuration data, you must take this into account when planning a cold standby scenario. For this scenario, you must use a full SQL server.

For this cold standby setup, you may need a second server for your SQL server to run on. To prepare for the cold standby, install Windows and BES. Ensure that you use the same revision as the original BES (this includes all hot fixes and service packs). When setting up this second BES, ensure that you configure it to point to the second SQL server and not the same one that the primary BES is using. You must also use the same SRP ID and SRP Authentication Key. To be safe, change the server document so that no users can access the server. Finally, you must have a copy of the primary Domino server's server.id file.

Once this has been completed, you must set up one-way Domino replication of the databases found in the BES directory, including the State Databases found in the \state subdirectory. In addition to the state databases, you must replicate the BlackBerry Profiles database as it contains the user's encryption keys. You must remember that whenever you add a new user to the BES, you must manually set up replication for the new State Database.

The last step is to set up one-way SQL replication of the Configuration Database (BESMgmt) to the SQL server in your disaster-recovery site.

In the event of a disaster, change the server document to allow access by all users and start the BES task on the secondary server. Users should not notice any interruptions unless there was user activity between the last replication cycle and the server going down. If that is the case, some users may have one or two e-mails on their handheld devices that they will not be able to reply to or forward.

After you have switched over to the secondary server, you must address the use of the BlackBerry Desktop Manager in your environment. There are two ways to handle this. You can manually switch the server setting in the Desktop Manager to the standby server at each user's desktop, or you can use the primary server's server ID file on the secondary server. This enables the Desktop Manger to continue working after a restart, since it is still accessing the same Domino server.

One important thing to remember is that if the BES that failed is an MDS push server, applications that you have written to push data to the handheld devices must be modified to point to the new server name, since the Windows server name will be different. One way around this is to set up a BES in the disaster-recovery site that is your only MDS push server. This way, in the event of a disaster, all MDS push functionality will continue to work. This is important because you may be using MDS to send out important information during a disaster.

When it is time to switch back to the primary BES, you must bring up the primary server off the network and disable the BES task. If you used the primary server ID file on the secondary server, you must stop the BES task on the secondary server, and then switch the ID back to the secondary server ID.

Connect the primary server to the network and quit the BES task on the secondary server. Change the server document to deny all user access, and begin replicating the data from the BES databases back to the primary server. Next, do a one-way replication of the BESMgmt SQL database so that the primary SQL server is up-to-date. When both primary Domino and SQL servers are up-to-date, change the primary Domino server's document to allow access by all users and start the BES task.

Knife-Edge Cutover

To achieve a knife-edge cutover, you must use a Domino cluster and an SQL cluster. To begin, install two SQL servers in a cluster so that the actual Configuration Database (BESMgmt) is a clustered database and not tied to a single SQL server. Since your BES is communicating with an SQL database with a virtual location, you will not need to reconfigure anything when you cut over. You will probably want to plan this kind of disaster-recovery implementation before you install your first 4.0 BES.

Install a second Domino server with the same software revision as the primary. Install the same version of BES software (including service packs and hot fixes) onto that server. Ensure that the BES task does not start automatically by removing it from the `notes.ini` file. In addition, set all of the BlackBerry services to manual and ensure that they are stopped. Modify the server documents so that all users are denied access to this server.

Next, create a cluster and place both primary and secondary servers into the cluster. Ensure that cluster replication is enabled so that all the changes in the BES databases are instantly replicated to the secondary server.

In the event of a disaster, start all of the BlackBerry services on the secondary Domino server and start the BES add-in task. Since the BESMgmt SQL database is virtual, it will remain available. Change the server document on the secondary server to allow access by all users. Change the server document on the primary server to deny access to all users.

All BlackBerry functionality will work normally with the exception of MDS, if this server was an MDS push server. Any push applications must be modified to point to the new Windows server name.

When it is time to switch back to the primary server, start the primary server off the network and quit the BES task. Connect the primary server to the network and quit the BES task on the secondary server, plus all BlackBerry services. Allow the replication to complete and start the BES task on the primary server.

Moving Users

Because all user configuration data is stored in the SQL Configuration Database, you can move users between BES servers if both BES servers are using the same Configuration Database. To move a user from one BES to another while both BES servers are running is a case of drag and drop. The source BES replicates the user's State Database to the destination BES, and the Configuration Database is updated. The destination BES sends out a new Service Book to the user's handheld unit so that it now communicates with the destination BES.

We can use this to our advantage when planning for a disaster. If you installed a fully functioning BES in your disaster-recovery location with no users on it, it could be used in the event of a disaster. There are a few things that must be in place before this can occur.

First, you cannot move a Domino BES user from a source to a destination BES without a State Database. In the event of a disaster, it is very likely the primary BES will be down, and so the State Database will not be available. You can plan ahead by manually replicating all State Databases and the BlackBerry Profiles database from the primary to the backup BES. If you move a user from one BES to another, when the source BES is down, the move will still occur if the user's State Database is already on the destination BES.

In addition to replicating the State Databases, you must use an SQL database that is still up. To achieve this, you can point both primary and secondary BES servers to one SQL database in your disaster-recovery location. That way, when the primary location goes down, you still have full control over your BES.

In the event of a disaster, all you need to do is drag and drop all of your users to the second BES. The great thing about this scenario is that when your primary location comes back up, you can simply drag and drop all of your users back to the primary BES.

Exchange BES

The Exchange BES has evolved ahead of the Domino BES over the years and the differences between version 3.6 and 4.0 in the back end are small. Both versions of BES use MSDE or SQL to store the configuration data, both versions of BES use MAPI, and BlackBerry data is still stored in hidden folders within each user's mail file. With this in mind, let's examine some disaster-recovery scenarios, including the following:

❑ Backup and restore

❑ Knife-edge cutover

❑ Moving users

Backup and Restore

If you choose to rely on restoring data from tape to reinstate your BES, you must ensure that you back up the following data:

❑ **The BlackBerry Service Account:** This is what the BlackBerry Service uses to run.

❑ **The BlackBerry Service Account Mailbox:** This contains critical information that allows the BES to start up and run correctly. It also provides initial MAPI connectivity for the BES, performs the address book lookups, and performs name/address resolution.

❑ **The SQL database:** This contains the BES and user Configuration Database (BESMgmt).

❑ **All BlackBerry** *users' Exchange mailboxes*: These must be backed up because they contain BlackBerry-specific settings in hidden folders.

In the event of server downtime, build a new BES server (and new SQL server if you used a separate server for this), restore all of the previously mentioned data, and start the BES. If the Exchange server is also down, you must restore the user's mailboxes so that their BlackBerry devices will continue to function.

Knife-Edge Cutover

To employ a knife-edge cutover disaster-recovery scenario, you must only have one BES instance on your server if you are using BES 4.0. Build a second server to act as your standby server that has the BES installed using the same SRP ID, SRP Authentication Key, and BES server name. The name of the Windows server (NetBIOS name) and the IP address of this second server need to be different from those of the primary server. The MAPI profile does not need to be the same. This cold standby server will have all BlackBerry services stopped.

Both the primary and standby BES are configured to communicate with the same SQL database (BESMgmt). The standby server could be pointed to a replicated copy of the BESMgmt database.

When the primary server goes down, simply start the BlackBerry service on the standby server. The BlackBerry users will see no interruption in service. When the primary server is ready to be brought up again, shut down the BlackBerry services on the standby server and do a one-time, one-way replication of the SQL data back to the primary SQL server. Once that has been completed, start the primary BlackBerry server.

User Move

The idea of this scenario is to keep two BlackBerry servers running, both sharing the same SQL database (BESMgmt), and each supporting its own BlackBerry users. Each BlackBerry server must have the capacity to support double the users that it currently supports. This is so that in the event that one of the servers becomes unavailable, the users from the disabled BES can be moved to the active BES. When the disabled BES is back in operation, the users can be moved back.

This scenario assumes that your handheld devices are all running the 4.0 code so that the service books can be updated wirelessly to point the handheld unit to the new server. To provide further redundancy, you can use a clustered SQL database.

Novell GroupWise BES 4.0

As of this writing, there are no documented disaster-recovery methods for the Novell GroupWise BES environment. It is likely that RIM will soon publish a white paper or knowledgebase document on how to achieve this.

Summary

Disaster-recovery planning is an important factor to consider when setting up any IT infrastructure or system. Since BlackBerry devices have become such an important business tool, keeping the data flowing to and from your BlackBerry devices during a disaster is critical.

As discussed in this chapter, you can make use of a plan that is as simple as backup and restore, or as complex as the knife-edge cutover scenarios that allow zero downtime. What you decide to implement will depend on how critical BlackBerry devices are in your environment, and how much your company is willing to spend on this.

In Chapter 9, we will discuss Mobile Data Service in depth. We will cover the way MDS works and the way in which it can be used within your organization to extend the capabilities of the BlackBerry architecture.

Part II

Expanding the Reach of Your BlackBerry Environment

Introduction to Mobile Data Service and Simulators

Mobile Data Service (MDS) is a secure conduit between your BlackBerry and your BES. Making use of this conduit allows you to extend your intranet services to your BlackBerry users, no matter where in the world they are. This chapter explores the different ways in which you can quickly and easily deploy enterprise-ready applications to your BlackBerry user population. This ultimately enables them to do more work in the field. Because the next few chapters make use of the BlackBerry simulators extensively, we will discuss them here.

Using MDS

To begin using MDS, you must first ensure that you have it installed in your environment. The most efficient way to install MDS is to install it onto every BES that you have in your organization, and assign one BES to be the MDS Push Server. The advantage of assigning one MDS Push Server is that when you develop applications that you want to "push" out to your BlackBerry users, you only have to know about one BES. That one MDS Push Server will take care of pushing your application out to all of your BlackBerry users, no matter which BlackBerry server they are on.

MDS runs as a standalone Windows NT service, but your Lotus Domino, Microsoft Exchange, or Novell GroupWise BES communicates with it extensively as part of its normal operation. In fact, the data sent between the MDS and the BlackBerry must go through the BES, because it is the BES that communicates directly with the BlackBerry NOC.

In addition to providing a secure conduit between the handheld unit and the BES, MDS has special functionality to deal with HTTP traffic. Because the BlackBerry handheld browser does not have the full functionality of a desktop browser, MDS must do some preprocessing on the data before it is passed on to the handheld devices. It also scales and optimizes any images before they are sent on to the handheld units, so that they fit nicely into the screen dimensions. In addition to processing images, MDS transcodes incoming HTML code before sending it on to the handheld devices. (*Transcoding* is a process of converting data from one format to another.)

Because of limited memory in the handheld device, MDS acts as a cookie repository for each handheld unit, to ensure that any cookies that are needed when returning to certain Web pages are honored.

MDS also supports certain special commands that enable you to push content to the handheld device. Following are these commands:

❑ **Browser Message:** This is a method of pushing a Web page to the handheld device as a message so that it appears in the Messages application.

❑ **Browser Content:** This is a method of pushing a Web page to the BlackBerry Browser cache. It can also be used to update a page that is already in the browser's cache.

❑ **Browser Channel:** This is a method of pushing a Web page plus two icons to the handheld unit. These three components make up the channel, and one of the icons will appear on the Home screen of the handheld device. The two icons represent a *read* and *unread* state. If the channel is refreshed, the icon will change status to unread until the user clicks it, and then it will change to the read status.

These three MDS functions are discussed in greater detail in Chapters 10 and 11.

Out-of-the-Box Usage

After you have MDS installed and enabled for all BlackBerry users, they will immediately be able to start using it. At this point, you do not need any knowledge of how MDS works, or how it responds to special commands. All you need to know is that it is providing a secure conduit between your internal network and the handheld devices. This out-of-the-box configuration enables you to use it in a number of ways.

Web Browsing (Internet and Intranet)

When you enable MDS for a BlackBerry user, a new icon (called the BlackBerry Browser) appears on the BlackBerry device. In pre-4.0 BlackBerry environments, this icon only appears after the user has synchronized the BlackBerry to the Desktop Manager. In 4.0 environments, the icon appears after the Service Book has been automatically pushed out wirelessly. The BlackBerry Browser application allows the BlackBerry user to browse any Web sites that can be accessed by their BlackBerry server. This normally means that users can browse to intranet sites and external Web sites on the Internet.

Your organization most likely has a firewall with content-filtering software on it that will block access to certain Web sites. This will mean that the user's BlackBerry Browser experience will be the same as it is when using a Web browser from a desktop in the office. Of course, the browser on the BlackBerry has limitations. When you browse Web sites that were not designed for portable devices, the formatting may look different and navigating them may be more difficult. On its own, the BlackBerry Browser is a very useful tool that enables you to offer Web browsing to your BlackBerry user population, while still controlling what users have access to, based on your organization's rules and regulations.

You are probably asking yourself why you would offer this feature to your users. You already may know that there is a way that your users can browse the Internet from their BlackBerrys by using the cellular carrier provider's browser icon. For example, T-Mobile has the t-zones icon and AT&T has the mMode icon.

A few factors make this option limiting or unavailable. The first drawback of the cellular carrier–provided browser is that when you use the browser, you are limited to browsing the Internet through a Wireless Access Protocol (WAP) gateway. The carrier may block many sites at its WAP gateway, even if the sites are WAP compliant. The carrier may certainly block non-WAP Web sites.

The second problem is that not all cellular carriers enable you to browse from your BlackBerry devices, which means that the option is not even available to the users.

Remember that your BlackBerry users will be able to browse your intranet. This is important because with just a little user education, you can create a BlackBerry portal for your users and just tell them what the URL is. From a BlackBerry Browser, the users must choose Go To and type in the URL. Users will instantly have access to your portal no matter where they are.

All you need to do from that point onward is ensure that any content you publish to that intranet site is easily viewable on the BlackBerry. We will cover how to achieve this in greater detail in Chapter 10.

Figure 9-1 shows the BlackBerry Browser displaying a Web site that has been designed with dimensions of a mobile devices screen in mind, plus the limited capabilities of the device.

Figure 9-1: Web site designed for a mobile device

Figure 9-2 shows the regular desktop browser version of that same Web site, while Figure 9-3 shows how the regular version is displayed on the BlackBerry. You can see that MDS has done as much as possible to the content of the page so that it looks good on the handheld devices and is quite usable. This example shows that browsing regular non-PDA friendly Web sites is possible on the BlackBerry, although maybe sometimes not as practical as a site designed specifically for mobile devices.

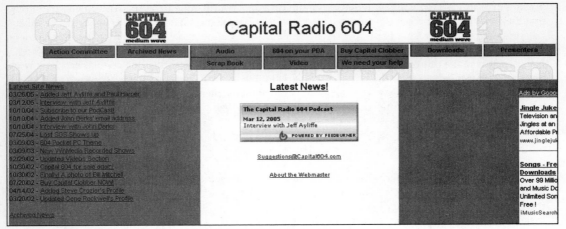

Figure 9-2: Desktop Web browser view of a regular Web site

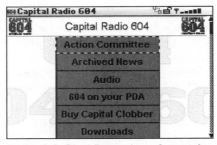

Figure 9-3: BlackBerry view of a regular
Web site

Playing Games

You may or may not allow your users to download and install software onto their BlackBerrys, but if you do, then they will be able to play games. Many of the games use MDS to send your high score to a central high-score server on the Internet so that you can see your ranking for that particular game. Figure 9-4 shows an example of such a server.

Figure 9-4: Screen shot of Raging Rivers
from MagMic

Some games enable you to play live with other players. This adds that extra enjoyment to the gaming experience and the thrill of being able to play with any BlackBerry user anywhere on the planet.

These games are written to take advantage of MDS, since it already provides a conduit to each BlackBerry and, therefore, the creators of these games do not have to write their own code to provide this functionality.

Off-the-Shelf Applications

Many software vendors provide off-the-shelf applications that make use of MDS. Many of these applications require additional setup and server resources on your part before they can function, but the payoff is large once they are installed.

There are many categories of applications, and each category of application requires more or less work to implement. There are too many applications and categories to mention, but here are a few to give you an idea of what is available.

Network Management

Network Management applications enable system administrators to manage many servers and applications on their networks. Products such as Idokorro (see Figure 9-5) and Sonic Admin (see Figure 9-6) enable you to manage Windows, Novell, Lotus Domino, BlackBerry, UNIX/Linux, and SQL Servers from your BlackBerry. You can also use Telnet or Secure Shell (SSH) to access any switch, router, or any other Telnet or SSH node on your network. These products also provide access to Novell eDirectory and Microsoft Active Directory. This enables you to manage all of these systems from anywhere on the planet where cellular coverage is available through your BlackBerry.

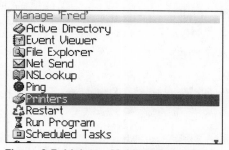

Figure 9-5: Idokorro Mobile Admin client

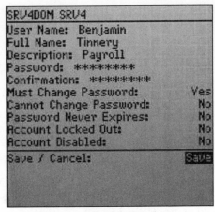

Figure 9-6: Sonic Admin client

Extending Your Intranet

Accessing Document Management systems such as iManage enables users to search documents they are working on and have them delivered to their BlackBerry devices. Products such as METAmessage, shown in Figure 9-7, make this possible. They work by using a client on the BlackBerry and XML forms to allow for searching and displaying results, while communicating to the back-end services through SQL and program-specific APIs.

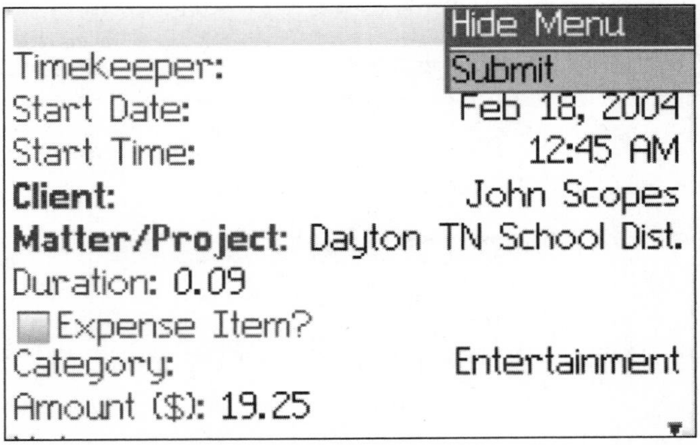

Figure 9-7: Typical METAmessage form

Another kind of product that falls into this category is Time Management. These products also have BlackBerry client and server components. They allow users who need to bill out their time to enter quickly and easily client and matter information, plus any relevant details on a meeting or appointment. This information will be inserted into a back-end system that is tied into the billing system. This is beneficial for people who are unreliable at keeping track of their own hours, or who do not want to have to go back to the office or boot up their laptops to enter their hours. Examples of these kinds of applications include METAmessage and TimeKM.

Handheld-Only Applications

BlackBerry client-side software can function without the need for you to set up any server software on your intranet. These kinds of programs communicate with outside Internet servers. An example of one of these applications is a program that monitors the stock market.

All of these applications can exist because of MDS. It is MDS that enables them to communicate over the Internet, or with their servers on your Intranet.

Introduction to the Simulators

When you begin creating content for your handheld devices you will want to know how your work looks. While you can imagine what it will look like, there is nothing better than seeing it firsthand. If it were Web content, you could use your BlackBerry to browse to the Web page. If it were a Java application, you could install the new code over the air and run it on the handheld unit.

The BlackBerry Simulator provides an almost perfect representation of the real BlackBerry handheld device, but running on a Windows computer. This enables you to stay at your desk, develop the handheld content, and see what it looks like on a computer. The BlackBerry Simulator is installed as part of the BlackBerry Java Development Environment (JDE). When you install the BlackBerry JDE, you get the BlackBerry Simulator, plus an MDS Simulator, and an Email Server Simulator. The MDS Simulator provides the functionality of a real MDS service and enables the BlackBerry Simulator to send and receive MDS content. The Email Server Simulator provides the functionality of a real mail server, which enables the BlackBerry Simulator to receive actual e-mails. Combined, these components help you see your projects look and function in the real world before you even use a real BlackBerry handheld device.

In addition to helping you develop applications, the Simulators can be very useful when creating documentation for your users, or for internal training classes. The BlackBerry Simulator, for example, enables you to take a screen shot for the entire simulated BlackBerry, or just the Liquid Crystal Display (LCD) screen.

BlackBerry Simulator

The BlackBerry Simulator is the program that includes the look and feel of each handheld model: how the device interacts with the data and voice network, how it interacts with your computer via a USB cable (or serial cable), and how the device reacts to varying signal strength and battery life.

Starting the Simulator

Before we begin, download and install the BlackBerry JDE 4.0. You can download it from RIM's Web site at this URL:

```
https://www.blackberry.com/SoftwareDownload/index.jsp?client=HPSHRfbRP&prod=⊃
1192&code=1758
```

After you have installed the BlackBerry JDE 4.0, you can start the BlackBerry Simulator by clicking Start ➪ All Programs ➪ Research In Motion ➪ Blackberry Java Development Environment 4.0 ➪ Device Simulator. When you select this application, a batch file is run that starts up the BlackBerry Simulator to simulate a BlackBerry 7290 (see Figure 9-8). If you want to simulate different models of the BlackBerry, browse to the `c:\program files\research in motion\blackberry jde 4.0\simulator` directory. There you will see batch files representing many different BlackBerry models. Each batch file is actually running a program called `fledge.exe`. This is the main simulator executable and it has a long list of command-line switches that enable you to control which handheld it simulates, plus many other characteristics of the simulated experience.

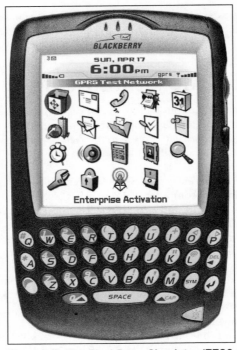

Figure 9-8: The BlackBerry Simulator (7730 shown)

Let's examine one of these batch files and see the very basic command-line options.

```
@echo off
..\bin\fledge /app=Jvm.dll /handheld=7290 /app-param=DisableRegistration
/app-param=JvmAlxConfigFile:7290.xml /data-port=0x4d44 /data-port=0x4d4e
/pin=0x21000004
```

Following are some of the options for this command:

❑ `@echo off`: This batch-file command simply instructs the command window not to show any text. The next line contains the actual command that starts up the simulator.

❑ `..\bin\fledge`: Instructs the batch file to move up one directory, and then down into the BIN directory where the `fledge.exe` program resides.

❑ `/app=Jvm.dll`: This switch tells `fledge` what application to run in the simulator. For our purposes, this will always be the Java Virtual Machine (`jvm.dll`).

❑ `/handheld=7290`: This tells `fledge` which handheld to simulate. Based on this information, it knows which handheld operating system to load, which skin to use, and the screen properties and dimensions.

❑ `/app-param=DisableRegistration`: Tells the simulator that when the BlackBerry starts up, not to attempt to register on the wireless network.

❑ `/app-param=JvmAlxConfigFile:7290.xml`: This tells the simulator what BlackBerry applications to include or exclude when the handheld device starts up. If you leave this parameter out, the simulator will load up every BlackBerry application that it finds in `c:\program files\ research in motion\blackberry jde 4.0\simulator` directory. If you want the simulated BlackBerry to load third-party applications, copy the `.cod` files from those applications to this directory.

If you want to have more control over which applications are loaded and which are not, you must edit the XML file for that model of BlackBerry. Take a look at the XML file for the BlackBerry 7290 (`7290.xml`):

```
<SimulatorConfiguration version="1.0" hardwareid="0x9C000503" flashSize="16384"
platformVersion="1.8.0.0">
    <ALXSources>.</ALXSources>
    <Application>net.rim.java.tasks</Application>
    <Application>net.rim.java.memopad</Application>
    <Application>net.rim.java.docViewer</Application>
    <Application>net.rim.java.phone</Application>
    <Application>net.rim.java.browser</Application>
    <Application>net.rim.java.browser.wtls</Application>
    <Application>net.rim.java.browser.ssl</Application>
    <Application>net.rim.java.browser.javascript</Application>
    <ApplicationExclude>net.rim.theme.vodafone</ApplicationExclude>
    <ApplicationExclude>net.rim.theme.tmobile</ApplicationExclude>
    <ApplicationExclude>net.rim.BBXpEnabler</ApplicationExclude>
    <ApplicationExclude>net.rim.blackberry.lang.chinese</ApplicationExclude>
</SimulatorConfiguration>
```

Using the `Application` and `ApplicaitonExclude` XML enclosures, you can include or exclude certain applications.

After the simulator has started, use the scroll wheel on your mouse to control the scroll wheel on the simulated BlackBerry. The right mouse button is mapped to the Escape button on the BlackBerry (the one below the scroll wheel). The computer's keyboard is mapped to the simulated BlackBerry's keyboard. The Control (Ctrl) key on your computer's keyboard is mapped to the simulated BlackBerry's Control key (half moon key). The two Shift keys are mapped to the simulated BlackBerry's two Numeric keys (used to select a number on a multipurpose key).

Simulating a Phone Call

To simulate how the BlackBerry will react to an incoming call, select the Simulate menu and choose Incoming Voice call. A new screen pops up, as shown in Figure 9-9. The top field enables you to enter the phone number of the supposed caller. The next field enables you to choose whether that caller is sending caller ID information, whether it is private, or whether it is unknown. After you have entered the phone number, click Create Call. The simulated BlackBerry will begin playing the ring tone.

When you answer the call on the simulated BlackBerry, the pop-up screen will change, as shown in Figure 9-10. It will show a drop-down menu that enables you to choose some kind of call failure. After you have chosen the type of call failure, click the Fail button to see how the BlackBerry handles that failure condition. To disconnect the call from the caller's side, click Disconnect.

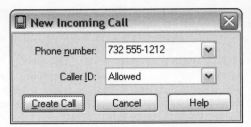

Figure 9-9: Simulating an incoming phone call

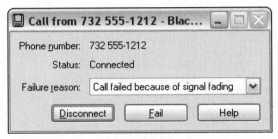

Figure 9-10: Simulating a failed call

Simulating Battery Conditions

To simulate how the simulated BlackBerry behaves under certain battery conditions, click the Simulate menu and choose Battery Properties. When the new screen pops up, you will see that you can simulate certain battery conditions, including a dead battery. Of course, when you select a dead battery, the hand-held unit displays a message and shuts down.

Simulating Signal Strength

To simulate how the simulated BlackBerry performs under different signal strength conditions, click the Simulate menu and choose Network Coverage. The screen that pops up shows a sliding bar that enables you to adjust the Received Signal Strength Indicator (RSSI) value.

Connecting the Simulated BlackBerry to a Real BES

You can connect a simulated BlackBerry to a real BES. To do this, first turn off the radio by using the simulated handheld. Then, click the Simulate menu and choose USB Cable Connected. If you have a HandHeld Manager or Desktop Manager running on the computer, the simulated BlackBerry will be recognized.

Allow the Desktop Manager to switch the PIN to the PIN of the simulated BlackBerry, generate the encryption key, and allow it to go through the Enterprise Activation. After it has done that, the simulated BlackBerry will be live on the BES and assigned to a real e-mail user. You will now be able to use the simulated BlackBerry just like a real one, including making use of MDS, e-mail, calendar synchronizing, and so on. Of course, if you just want to simulate MDS, you do not need to connect your simulated BlackBerry to a real BES. All you need to do is start the MDS Simulator.

MDS Simulator

The purpose of the MDS Simulator is to simulate the MDS for the BlackBerry Simulator. This enables the simulated BlackBerry to access all MDS services and enables you to test any MDS-dependent applications you may have built.

Some MDS commands enable you to push content to the handheld devices by using the user's e-mail address. If you were using a real BES and real MDS service, the MDS service would know which handheld to send the data to based on a table it keeps that matches the e-mail address to the handheld PIN.

If you want to test such push commands using the simulated BlackBerry and the MDS Simulator, you first need to edit a file. Browse to `c:\program files\research in motion\blackberry jde 4.0\mds\config` and edit the file called `rimpublic.property`. Scroll down through this file until you see the heading `# [Simulator]`. Under this heading you will see a list of PINs and e-mail addresses:

```
# [Simulator]
#SimulatorMDS.TEST=localhost:8081
Simulator.2100000a=MDS,simulator@pushme.com
Simulator.2100000b=MDS,user2100000b@pushme.com
Simulator.2100000c=MDS,user2100000c@pushme.com
Simulator.2100000d=MDS,user2100000d@pushme.com
Simulator.2100000e=MDS,user2100000e@pushme.com
Simulator.2100000f=MDS,user2100000f@pushme.com
```

If you want to test a push command using the BlackBerry Simulator, first ensure that when it starts up, it is using the correct PIN. When you issue the MDS push command, send it to the correct e-mail address. We will discuss MDS push commands in more depth in Chapters 11 and 12.

Email Server Simulator

If you must test an application that uses e-mail, or you just want to see how your application behaves when the BlackBerry receives e-mail, you can use the Email Server Simulator (see Figure 9-11). The Email Server Simulator will collect e-mail on a real mail server using POP3 and send e-mail using a real SMTP server. E-mail received from a POP3 server will be delivered to the handheld. E-mail sent from the handheld will be sent to the SMPT server.

There is currently a bug with the Email Server Simulator that you can easily correct. You need to do the following to allow it to function correctly:

1. Edit the file called `rimpublic.property`; you'll find it in `c:\program files\research in motion\blackberry jde 4.0\ess\config`.

2. Add the line **Email.mode=Exchange** at the bottom of the file.

This does not actually tell the Email Server Simulator to use Exchange, but there must be a value for this parameter before the Email Server Simulator will work correctly. You can put any value in there and it will have the same effect.

If you would like to send and receive e-mail using your internal e-mail system, simply enable POP3 and SMTP access to your e-mail environment, and use your regular login information in the Email Server Simulator.

Figure 9-11: Email Server Simulator Configuration Screen

Summary

So far, you can see that by the simple act of enabling MDS, you will open a whole new set of functionality for your BlackBerry environment. You can offer Web browsing if your cellular carrier does not. You can create a BlackBerry portal intranet site for your users to connect to while on the road, and you can make use of off-the-shelf applications to make internal company data available to your users.

There is much more that can be done with MDS, and we will cover these topics in Chapters 11 and 12.

In Chapter 10, we examine how to set up a BlackBerry Web portal. We will discuss ways in which you can optimize your Web content to cater to BlackBerrys. We will have a short tutorial on WML, which is a language written specifically for mobile devices.

10

BlackBerry Web Portal

You can offer company information to your BlackBerry users by building an intranet portal. Of course, you may already have an intranet portal designed to serve up internal company information to your network users. A portal designed for your BlackBerry users, however, will be special.

A BlackBerry portal can not only serve up internal company information, it can also enable your BlackBerry users to gain access to data that will help them while they are traveling and visiting clients. This information will be useful and beneficial to their jobs and will enable them to be more productive.

Examples of useful applications would be the ability to access Contact Resource Management (CRM) data, the ability to enter time and billing information directly into your company's time and billing system, the ability to update or modify help desk tickets while out in the field, and the ability to interact with other internal Web systems without having to boot your laptop and find a connection to the Internet.

This chapter explores the ways in which you can build a BlackBerry-friendly Web portal within the context of a company Cafeteria Portal.

The sample code used in this chapter can be downloaded from the Wrox Web site. The name of the file is `589539 ch10 code.zip`.

Creating BlackBerry-Friendly Web Content

The BlackBerry has limitations that prevent it from displaying Web pages in the way that the pages display on a desktop computer screen. For example, there are limitations to the screen size. The BlackBerry screens are either monochrome or color, and they range in size from 160×160 pixels to 240×240 pixels. A typical desktop computer screen is 1024×768 pixels or larger.

In the case of the monochrome BlackBerry handhelds, the screen is true black and white and not grayscale. This means that you must design the images so they look good with only two colors. The amount of screen real estate that you have on a BlackBerry is much smaller than on a regular desktop computer. So, when you design your Web content, you must accommodate these smaller screens.

In addition to screen size and number of colors, you must be aware that the available bandwidth to these devices is much lower than a broadband-connected or LAN-connected computer. The bandwidth to a BlackBerry is equivalent to dial-up speeds (sometimes less, depending on signal strength).

These limitations are not as bad as you might think, because the MDS helps out in a very big way. It transcodes the data between your network and the BlackBerry so that the content is much smaller. It reduces the size of images so that they fit more easily on the BlackBerry screen.

Depending on the device being used to browse, MDS uses the following methods to transcode images:

❑ **Horizontal scaling:** If an image on a Web page is larger than the screen dimensions, then MDS will reduce the size of the image to no larger than the width of the screen minus 5 pixels. MDS reduces the image's width by the extra 5 pixels because the scrollbar takes up 5 pixels on the right side of the screen. When the images are reduced, their aspect ratio is maintained so that the length is reduced in proportion to the width reduction.

❑ **Vertical scaling:** If an image on a Web page is longer than twice the screen length, then MDS will reduce it so that its length is twice the length of the device's screen. When the length is reduced, the width is reduced in proportion so that the images original aspect ratio is maintained.

❑ **Monochrome size limitations:** After monochrome images are reduced, if their size is more than 8,192 bytes, then the image is discarded. If the image was part of a link, the link's alternative text will be used in place of the image.

❑ **Color images:** Color images are dithered into monochrome images for display on a monochrome BlackBerry. The number of colors in a color image are also reduced, if necessary to conform with the number of colors that the color BlackBerry can display (65,000).

❑ **Vertical alignment:** Vertical alignment in <img> tags of the Web page's HTML code are ignored.

MDS strips out any HTML code that it knows the handheld unit will not understand (this is to reduce the size of the Web pages). This means that you do not necessarily need to create your BlackBerry portal with Wireless Markup Language (WML), unless you have non-Java handheld devices that require access to the portal. If you do have non-Java handheld devices (such as the Motient 850 and 857, or the Cingular 950 and 957), you must create some WML content.

Because you may have different types of handheld devices in your user population, you need to accommodate the different screen properties and Web page languages by creating different versions of the portal based on the handheld device being used. You may even want to create a server-side script that detects the type of browser being used and redirect the device to the appropriate page. You could combine the handheld devices and regular portal on the same Web server by using this method.

Designing the Cafeteria Portal

To demonstrate and discuss the ways in which you can design a BlackBerry-friendly Web portal, we will do so in the context of a fictitious Cafeteria Portal.

Before we begin, let us compare what Web pages look like on a regular desktop Web browser and what they look like on the BlackBerry Web browser. To do this, launch the MDS Simulator. When it is running, start the Device Simulator. We will begin with the BlackBerry 7290.

Using the simulated BlackBerry 7290, open the BlackBerry Browser, click the scroll wheel, choose Go To, and type a URL for a regular Web site (for example, `engadget.com`, `cnn.com`, or `msnbc.com`). Use a Web site that does not cater to mobile device browsing. Open that same Web page in your favorite desktop Web browser. Now take a look at the different ways in which the Web site is rendered. On the desktop browser, everything appears as it should. On the simulated BlackBerry 7290, you will notice that the content is all there with the exception of a few changes. Any images on the Web page have been reduced in size. If frames are used on the Web site, they will not show on the BlackBerry.

As you use your simulated BlackBerry 7290 to browse these Web sites, you will discover that MDS does a great job in transcoding the Web content for the handheld unit, and the handheld device does a great job in rendering it in a way that still enables you to navigate and read the content. It also proves that you can design your internal Web portals with regular HTML.

Wireless Markup Language

If you have any older C++ based handhelds (such as the Cingular 950 and 957, or the Motient 850 or 857), then your portal will need to support WML. These older handhelds can only render WML code and not HTML, JavaScript, or CSS. WML is not a difficult language to learn, especially if you already know HTML. In addition to needing to use WML if you support these older handhelds, it is still a great way to build Web content that has a very small footprint and provides special functions that fit well in a world of wireless Web browsing. All of the BlackBerry handhelds currently support WML. Let us take a brief look at WML and some of the unique features it provides to the browsing experience.

If you would like to view the sample code on your own Web server, and haven't already done so, unzip the file called `589538 ch10 code.zip` and place the contents in the root of your Web server. If you do not have access to a Web server, then we recommend using the Abyss Web Server X1 from `www.aprelium.com`. After you have downloaded and installed the Abyss Web server, copy all of the files from `589539 ch10 code.zip` to `c:\program files\abyss web server\htdocs`.

Introduction to WML

The easiest way to visualize WML is to think of a deck of cards, as shown in Figure 10-1. One WML page is one deck, and that deck will contain one or many cards. All tags on a WML page must be in lowercase. Uppercase tags are illegal and will be ignored. All WML pages must start with the following header:

```
<?xml version="1.0"?>
<!DOCTYPE wml PUBLIC "-//WAPFORUM//DTD WML 1.2//EN"
"http://www.wapforum.org/DTD/wml_1.1.xml">
```

Last, all WML pages must start with `<wml>` and end with `</wml>`. Here is an example of a WML page:

```
<?xml version="1.0"?>
<!DOCTYPE wml PUBLIC "-//WAPFORUM//DTD WML 1.2//EN"
"http://www.wapforum.org/DTD/wml_1.1.xml">

<wml>
<card id="ProBB-Hello">
<p>
&quot;Hello, ... <br/>
... BlackBerry Guru!&quot;
</p>
</card>
</wml>
```

```
<?xml version="1.0"?>
<!DOCTYPE wml PUBLIC "-//WAPFORUM//DTD WML 1.2//EN"
"http://www.wapforum.org/DTD/wml_1.1.xml">
```

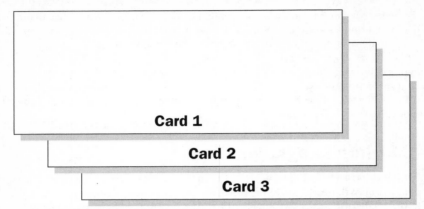

Figure 10-1: Visualization of a WML page (or deck) with multiple cards

The output of this page is shown in Figure 10-2 (if you load the MDS Simulator, the Device Simulator, and browse to `http://localhost/hello.wml`).

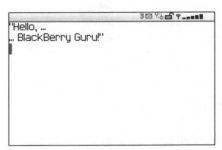

Figure 10-2: WML page on a BlackBerry handheld device

We are doing the following in this WML page:

❑ We give the card a name or identifier of `"ProBB-Hello"`.

❑ We then start a new paragraph with the <p> tag.

❑ We then write the phrase `"Hello, ...` on the first line. We have also used the double quotation marks (`&quot;`). In this example, we did not have to use `&quot`. We could simply have typed a double quotation character ("). We inserted this to demonstrate that you can use entities. We end the line with a line feed (`<br/>`).

❑ On the next line, we write the phrase `... BlackBerry Guru!"`. Again we use the double quotation marks (`&quot;`) at the end.

❑ The last two lines close the paragraph (`</p>`) and close the card (`</card>`).

Formatting

WML supports the usual formatting tags as listed here:

- ❑ Bold: `<b> </b>`
- ❑ Emphasis: `<em> </em>`
- ❑ Italics: `<i> </i>`
- ❑ Strong: `<strong> </strong>`
- ❑ Underline: `<u> </u>`

This next sample WML file demonstrates what each style looks like (see Figure 10-3). You can also load `http://localhost/style.wml` on your device simulator.

```
<?xml version="1.0"?>
<!DOCTYPE wml PUBLIC "-//WAPFORUM//DTD WML 1.2//EN"
"http://www.wapforum.org/DTD/wml_1.1.xml">

<wml>
 <card id="Styles" title="Styles">
  <p>
  This is regular text... <br/>
  This is <b>Bold</b>  text ... <br/>
  This is <em>Emphasized</em>  text ... <br/>
  This is <i>Italics</i>  text ... <br/>
  This is <strong>Strong</strong>  text ... <br/>
  This is <u>Underlined</u>  text ... <br/>
  </p>
 </card>
</wml>
```

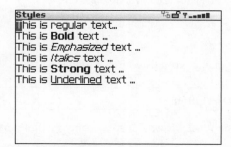

Figure 10-3: Appearance of styles

In this example, you will notice that we have placed a non-breaking space (` `) before the word `text`. We do this because, after certain WML tags, the browser will remove the next space character. To counteract this we insert a non-breaking space.

Alignment

WML supports aligned paragraphs. They can be left, right, or center. When you start a paragraph, you must specify the alignment at the beginning, as listed here:

- ❑ `<p align="left"></p>`

125

❏   `<p align="right"></p>`

❏   `<p align="center"></p>`

Figure 10-4 shows the following sample WML page as it looks on the handheld device. If you browse to `http://localhost/align.wml` on your simulated BlackBerry, you will be able to see it.

```
<?xml version="1.0"?>
<!DOCTYPE wml PUBLIC "-//WAPFORUM//DTD WML 1.2//EN"
"http://www.wapforum.org/DTD/wml_1.1.xml">

<wml>
<card id="Right-Align" title="Alignment">
<p align="right">
Each line<br />
of this text<br />
is aligned with the right of the screen.<br />
</p>
</card>
</wml>
```

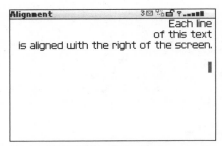

Figure 10-4: Right-aligned text

Displaying Images

The BlackBerry supports the image types listed in the following table.

Image Type	Description
GIF87 and GIF89	The WML browser displays only the first image of an animated GIF.
JPEG	JPEG images display only on a color BlackBerry unless the monochrome handheld device is set to use MDS. MDS will then convert the JPEG to a monochrome PNG.
PNG	By default, the browser converts all GIF images to PNG, because it has a much higher compression ratio and has support for alpha channels.
WBMP	These are Monochrome Wireless Bitmap (WBMP) images. They are low-resolution images.

Here is a sample WML page that displays a picture. Figure 10-5 shows how it looks on the handheld. To see it, browse to `http://localhost/pic.wml`.

```
<?xml version="1.0"?>
<!DOCTYPE wml PUBLIC "-//WAPFORUM//DTD WML 1.2//EN"
"http://www.wapforum.org/DTD/wml_1.1.xml">

<wml>
<card id="Pic" title="A Picture Demo">
<p align="center">
<img src="pic.png" alt="My Picture" />
<br />
This shows a picture.
<br />
</p>
</card>
</wml>
```

Figure 10-5: Image display

The BlackBerry WML browser does not support "built-in" images. These "built-in" images are pre-loaded images that are normally stored on a mobile telephone. The WML page can call them by name and, instead of them loading over the air, they pop right out of memory into the page. If you attempt to use one of these "built-in" images, the browser will just display the "alt" description.

Tables

WML supports tables with rows, columns, and formatting. You can place text, images, links, and style tags in table cells. You can format each cell in a table as either left-, center-, or right-aligned. However, the formatting command does not follow the same rules as regular text formatting. The Align tag is used and the letter L is used for Left, C for Center, and R for Right. In addition, you can specify the alignment of each column in one `align` statement, as shown here:

```
<table columns="2" align="CR"></table>
```

This command creates a table with two columns, and then aligns the first column as Center and the second as Right.

This sample WML page shows how to create a table, but also demonstrates how to format each cell in a row. You will notice that in the second row, the first cell is centered while the second is right-aligned. In the third row, the first cell is centered, the second is right-aligned, and the third is left-aligned.

```
<?xml version="1.0"?>
<!DOCTYPE wml PUBLIC "-//WAPFORUM//DTD WML 1.2//EN"
"http://www.wapforum.org/DTD/wml_1.1.xml">

<wml>
<card id="Table" title="Table">
<p>
<table columns="2">
<tr><td>R1 - C1</td></tr>
</table>
</p>
<p>
<table columns="2" align="CR">
<tr><td>R2 - C1</td><td>R2 - C2</td></tr>
</table>
</p>
<p>
<table columns="3" align="CRL">
<tr>
<td>R3 - C1</td>
<td>R3 - C2</td>
<td>R3 - C3</td>
</tr>
</table>
</p>
</card>
</wml>
```

Figure 10-6 shows how it will look on the handheld device.

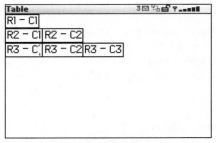

Figure 10-6: Handheld table view

Using Telephone Numbers in WML

If you use links to phone numbers and e-mail addresses, the browser will know what to do when you click the link. For example, if you click a phone number link, the BlackBerry will ask if you would like to dial that number.

Here is a sample WML page that lists work and home numbers and e-mail addresses. Open `http://localhost/phone.wml` in your simulator to see what it looks like.

```
<?xml version="1.0"?>
<!DOCTYPE wml PUBLIC "-//WAPFORUM//DTD WML 1.2//EN"
```

```
"http://www.wapforum.org/DTD/wml_1.1.xml">
<wml>
<card id="Contact" title="Contact: Arthur">
<p>
Call:
<a href="wtai://wp/mc;2125551212"
title="Call">Work</a>
<a href="wtai://wp/mc;2125551213"
title="Call">Home</a><br />
Email
<a href="mailto:adent@office.com">Work</a>
<a href="mailto:adent@home.com">Home</a>
</p>
</card>
</wml>
```

Multiple Cards in a Deck

All of the preceding examples have only one card in the deck. The power of WML is realized when you start using multiple cards in a deck and reference those cards so that the user can view what appear to be multiple pages without the browser having to load those pages.

Our sample WML page shows an example of a phone book. On the main page you will see two categories: Comedians and Musicians. When you click Comedians, you see a list of comedians with their associated phone numbers, Web sites, and e-mail addresses. At the bottom of the screen, you see the word Back. If you scroll down and click the word Back, the browser will go back to the previous card in the deck. That previous card happens to be the main page.

Let us discuss how this multicard deck is achieved. Figure 10-7 visually demonstrates this concept.

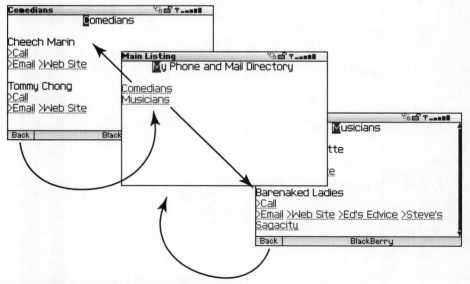

Figure 10-7: Visualization of a multicard deck

The first card looks like this:

```
<wml>
<card id="PhoneList" title="Main Listing">
<p align="center">
My Phone and Mail Directory
</p>
<p>
<a href="#Funny">Comedians</a>
<br />
<a href="#Tunes">Musicians</a>
</p>
</card>
```

This card is used as the main page and it has links (href) to two other cards. Links to other cards in the deck reference the card's ID. For example, to link to the card called Funny, we use this command:

```
<a href="#Funny">Comedians</a>
```

The next card in the deck is the Funny card (which lists the comedians), as shown here:

```
<card id="Funny" title="Comedians">
<p align="center">
Comedians
</p>
<p>
Cheech Marin
<br/>
<a href="wtai://wp/mc;5195551234">
&gt;Call
</a>
<br />
<a href="mailto:info@cheechandchong.com">
&gt;Email
</a>
<a href="http://www.cheechandchong.com">
&gt;Web Site
</a>
</p>
<p>
Tommy Chong
<br/>
<a href="wtai://wp/mc;5195551234">
&gt;Call</a><br />
<a href="mailto:info@cheechandchong.com">
&gt;Email</a>
<a href="http://www.cheechandchong.com">
&gt;Web Site</a>
<do type="prev" label="Back">
<prev/>
</do>
</p>
</card>
```

This card lists all the comedians and shows the word Back at the bottom of the screen. This virtual button is placed there by using the do command, which enables you to assign an action to a virtual button. These virtual buttons have their beginnings in the days when cellular telephone browsers were the only WML browsers. There were always these two physical buttons underneath the screen that could be used as virtual buttons.

So, in our example, we set the left buttons to "Back", which takes the user to the previous card in the deck when it is pressed:

```
<do type="prev" label="Back">
```

The predefined do types are accept, prev, help, reset, options, delete, and unknown. Here we used the type "prev" and labeled the button "Back".

Variables

You can set variables in WML pages by using the <setvar> and <input> tags. You can display the contents of a variable by preceding it with a $ sign.

This next sample WML page shows how variables can be used.

The first card called Variables sets up an Accept virtual button at the bottom of the screen with the label "Let's see". When it is clicked, the values that the user inputs into the form are accepted. In addition, the browser is taken to the card called "WhatIs".

Next, we set a variable called mListen and assign the value of "Depeche Mode" to it. We go on to set a variable called mAge and assign the value "37.34" to it.

```
<wml>
<card id="Variables" title="Variables Demo">
<p align="center">
Working with variables
</p>
<do type="accept" label="Lets see">
<go href="#WhatIs">
<setvar name="mListen"
value="Depeche Mode" />
<setvar name="mAge" value="37.34" />
<p>
```

Next, we set up a simple form asking you to input who you like listening to, your age, and your password. Here we are demonstrating that you can set the format of each input. For example, when we ask for your phone number, we input it into a variable called yTelephone, and we set the format as "\(NNN\)\NNN\-NNNN". You will notice that when you enter your phone number, the browser already knows that it should accept numbers and will format the phone number field correctly.

When we ask for your password, we input it into a variable called yPassword, and we set the maximum length to 12 characters and the format to "password". The format of "password" tells the browser to hide your input because it is sensitive information. This causes the browser to show each character you type as an asterisk.

```
You like to listen to?
<input name="yListen" />
```

```
What is your telephone number?
<input name="yTelephone"
format="\(NNN\)\ NNN\-NNNN" />
Your age?
<input name="yAge" size="2"
maxlength="2" format="NN" />
Your password?
<input name="yPassword" size="12"
maxlength="12" type="password"
format="*x" />
</p>
</go>
</do>
</card>
```

Last, we display all the variables in the next card, called `"WhatIs"`. Notice that we have to place the $ sign in front of a variable of which we want to display the contents. For example, we could type the following sentence:

```
I am $mAge years old and like to listen to $mListen<br />
```

When this is displayed, instead of $mAge, the browser will display the contents of the variable mAge.

```
<!-- display variables -->
<card id="WhatIs" title="Let us compare">
<p align="center">
My wonderful variables!
</p>
<p>
I am $mAge years old and like to listen to
$mListen<br />
You are $yAge years old and like to listen to
$yListen<br />
Your telephone number is $yTelephone<br />
Your "secret" password is $yPassword<br />
</p>
</card>
</wml>
```

To see what this looks like on a BlackBerry (see Figure 10-8), browse to `http://localhost/var.wml` on your simulator.

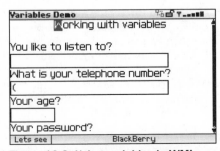

Figure 10-8: Using variables in WML

For a full list of WML commands and their syntax, please see Appendix A. There is much more to WML, and we suggest that you read an article called "WML 101" written by Richard Evers in the October 2004 issue of the *BlackBerry Developer Journal*. You will find it here:

```
www.blackberry.com/developers/journal/oct_2004/wml_101.shtml
```

Displaying the Correct Portal Page

Planning your new mobile Web portal is essential. There are a few things to consider before you start. One approach would be to use your company's existing Web portal and simply add new content that is useable on BlackBerry devices. You may choose to set up a totally new Web site that caters to the BlackBerry devices.

If you choose one of these two approaches, you may still need to make one more decision. If you choose to create a completely new Web site that only caters to your BlackBerry users, then you must decide if you want to design the Web pages so that they look and function the same, no matter what model of BlackBerry is used, or if you want to create one or two versions of each Web page to make the site look and work in the best way possible depending on which model of BlackBerry is used.

If you choose to make use of your company's existing Web portal, you must put something in place that redirects the user to the correct pages, depending on whether they are browsing using a desktop browser or a BlackBerry browser. You could extend this to redirect the user to the correct pages based on the model of BlackBerry being used.

The Redirector Page

The idea of a redirector page is a Web page that has some built-in logic to redirect a Web browser to another page based on certain criteria. In our example of the Cafeteria Portal, we will make the home page the redirector page. This idea is useful for any Web site on the Internet that publishes only one URL that is accessible from desktop computers and mobile devices.

There are a number of ways to achieve a redirector page, and they depend heavily on the kind of Web server you are using. For this example, we will assume that you are using a Microsoft IIS Web server that supports Active Server Pages (ASP). ASP is useful when you want the Web server to perform operations that the Web browser cannot necessarily perform. ASP is commonly referred to as a form of *server-side scripting*. When IIS finds ASP code in the Web pages, it does not send it to the Web browser, but rather processes it locally and takes whatever action is needed. Sometimes this is page redirection, sometimes it involves inserting data into the Web page.

For our redirector page, we will make use of a value that all Web browsers report: `http_user_agent`. The `http_user_agent` is a value that all browsers report to the Web server and it normally contains the kind of Web browser it is, including any version numbers. When a BlackBerry browser passes the `http_user_agent` value, it reports the actual BlackBerry model number and version of the RIM operating system it is running. This is extremely helpful to Web designers because it enables them to redirect the BlackBerry user to the correctly formatted Web pages.

Let us take a look at our Cafeteria Portal to see how we achieve this. You will be able to use these pages on your Web portal as a starting point and modify them to fit your environment.

The following `Default.asp` file is the home page of the Cafeteria Portal. This is what it looks like:

```
<html>
<head>
<meta name="GENERATOR" content="Microsoft FrontPage 5.0">
<meta name="ProgId" content="FrontPage.Editor.Document">
<META NAME="ROBOTS" CONTENT="NOINDEX, NOFOLLOW">
<meta http-equiv="Content-Type" content="text/html; charset=windows-1252">
<title>Browser Detect - Please wait</title>
<meta name="Microsoft Theme" content="none">
</head>
<body>
<%
    myUA = Request.ServerVariables("HTTP_USER_AGENT")

if Left(myUA,11) = "BlackBerry7" then
            Response.Redirect "BlackBerrycolor.asp"
      ElseIf Left(myUA,11) = "BlackBerry6" then
            Response.Redirect "BlackBerrybw.asp"
      Else
            Response.Redirect "PCPage.asp"
      End if
%>
</body>
</html>
```

If you have not already done so, download the file called `589539 ch10 code.zip` from the Wrox Web site and unzip the contents to the root of your Web server. If you do not have access to a Web server, as we mentioned before, we recommend using the Abyss Web Server X1 available at `www.aprelium.com`. After you have downloaded and installed the Abyss Web server, copy all of the files from `589539 ch10 code.zip` to `c:\program files\abyss web server\htdocs`. Make sure that you remove the original home page of the Web server. It may be called `index.htm` or `index.asp`. The only home page must be the one from the zip file, called `default.asp`.

The part of this Web page that does all of the work is the part between the percent signs about midway through the code. When you enable ASP on an IIS server, you can embed special ASP code into Web pages by putting it between percentage tags.

Let us go through the ASP code line-by-line to see how we are able to redirect the browser to the correctly formatted pages. First, look at the following lines:

```
<%
    myUA = Request.ServerVariables("HTTP_USER_AGENT")
```

Two things happen in this first line, and we will start with the second command first. When IIS sees the `Request.ServerVariables("HTTP_USER_AGENT")` code, it returns the browser information. When it is returned, we need to store it somewhere temporarily so that we can use that information for decision-making. In this line, we create a variable called `myUA` on the fly and we insert the data returned from the `Request.ServerVariables("HTTP_USER_AGENT")` command. This line creates a variable called `myUA` and inserts the browser information.

When the browser information is returned from a BlackBerry, it looks similar to this:

```
BlackBerry7290/4.0.0
BlackBerry7100/4.0.0
BlackBerry6230/4.0.0
```

The format of the returned browser information is `BlackBerry<model number>/<RIM OS version>`. So, a BlackBerry 7100 running RIM OS 4.0.0 would return `BlackBerry7100/4.0.0`

When RIM assigns model numbers to their BlackBerry devices, they use a specific numbering scheme. The first digit always represents whether the device is color or monochrome. A first digit of 7 tells you that the device is color, while a first digit of 6 tells you the device is monochrome.

Here are the rules for the model numbering scheme:

- ❑ **6xxx:** Monochrome
- ❑ **7xxx:** Color
- ❑ **x1xx:** Cellular phone form factor (e.g. 7100 series)
- ❑ **x2xx:** Small PDA form factor
- ❑ **x5xx:** Small PDA form factor with external antennae and Push To Talk (PTT)
- ❑ **x7xx:** Full PDA form factor
- ❑ **xx10:** World band (900 and 1900) or iDEN
- ❑ **xx20:** Dual band (900 and 1800)
- ❑ **xx30:** Tri band (900, 1800, 1900)
- ❑ **xx50:** CDMA
- ❑ **xx70:** WLAN
- ❑ **xx80:** Tri band (850, 1800, 1900)
- ❑ **xx90:** Quad band (850, 900, 1800, 1900)

The following are some exceptions:

- ❑ **5790:** Java Mobitex
- ❑ **58x0:** Named before conventions were in place, operates on the 1900 band
- ❑ **7100 series:** Letter following the first four digits represents the cellular carrier

For the Cafeteria Portal, we have decided to create a color and black-and-white version. We only need to concern ourselves with the first digit. With that in mind look at the second and third lines of ASP code:

```
if Left(myUA,11) = "BlackBerry7" then
            Response.Redirect "BlackBerrycolor.asp"
```

This is the beginning of the decision-making logic. We are using the information that is stored in the variable called myUA to make a decision on where to redirect the browser. When a BlackBerry returns its browser information, it is a known number of characters and a known character pattern. The first 10 characters will always be BlackBerry. The next character will be the first digit of the model number. So, for this redirection, we only need to be concerned about the first 11 characters of the browser information. This piece of logic tells IIS to start at the left of the data stored in the variable myUA and count across 11 characters. Then, take these first 11 characters and compare it with "BlackBerry7". If it matches, then redirect to a new Web page called "BlackBerrycolor.asp".

If the first 11 characters of the variable myUA do not match "BlackBerry7", then IIS moves to the next two lines of code:

```
ElseIf Left(myUA,11) = "BlackBerry6" then
            Response.Redirect "BlackBerrybw.asp"
```

The line of code starting with ElseIf tells IIS that if the previous If statement produced no matches, it must consider this new If statement. This statement is similar to the first one in that it tries to match the first 11 characters of the variable myUA, but this time it tries to match it to "BlackBerry6". If it does match then it redirects the browser to a new Web page called "BlackBerrybw.asp".

The last few lines of ASP code are there to deal with where to redirect the browser if the first 11 characters of the variable myUA do not match "BlackBerry7" or "BlackBerry6":

```
Else
            Response.Redirect "PCPage.asp"
      End if
%>
```

We do not really need to concern ourselves with the contents of myUA if they do not match "BlackBerry7" or "BlackBerry6", because it probably means that the browser is a desktop browser. If it is, we can simply redirect it to the desktop browser version of the portal. This is exactly what these lines do. They redirect the browser to "PCPage.asp".

If you wanted to interrogate the myUA variable further to possibly make a decision on redirecting the browser to a page designed specifically for Firefox or Internet Explorer (IE), then you could. Firefox returns a value similar to this:

```
"Mozilla/5.0 (Windows; U; Windows NT 5.1; en-US; rv:1.7.5) Gecko/20041107
Firefox/1.0"
```

IE returns a value similar to this:

```
"Mozilla/4.0 (compatible; MSIE 6.0; Windows NT 5.1; SV1; .NET CLR 1.1.4322)"
```

The Cafeteria Home Page

After the user's browser has been redirected to the correctly formatted page based on the user's BlackBerry (color or monochrome), or a desktop browser, the user will see the appropriate Cafeteria home page. You can test this by publishing all of the files in the `589538 ch10 code.zip` file to the root of your Web server. Make sure that you use a test Web server and not the main production Web portal. You can also install and use IIS or other Web servers on your PC. After you have the pages on your Web server, and enabled ASP on the Web server, launch the MDS Simulator and the device simulator of your choice. Use it to browse to the Web server.

As we mentioned before, if you do not have a local Web server, or one that you can use, we recommend using the Abyss Web Server X1 from `www.aprelium.com`. It is very fast and efficient and enables you to add third-party Common Gateway Interface (CGI) interpreters. You can download a third-party ASP interpreter from seliSoft at `www.selisoft.com` and a third-party PHP interpreter from `www.php.net/downloads.php`.

After you have installed the Abyss Web server, follow these instructions on how to install PHP and ASP support.

Adding ASP Support to the Abyss Web Server

After you have downloaded ActiveHTML from `www.selisoft.com` and installed it, you need to configure the Abyss Web server to use it.

Open Abyss Web Server's console, select Server Configuration, and click Advanced and select CGI Parameters. Ensure that CGI Processing Enabled is set to Yes.

Next, click Add in the CGI Interpreters table. In the Interpreter Path field, enter the path of the `ahtml .exe` executable file (for example, `c:\program files\selisoft\activehtml\ahtml.exe`). Enter **asp asa** in the Associated extensions, set Automatic Update of the CGI Paths (if available) to Yes, and press OK.

Figure 10-9 shows the Abyss Web server configuration after PHP and ASP support have been added. Remember to restart the Web server to activate the changes.

If you are using a local install of IIS or Abyss on your PC, then the URL would be:

```
http://localhost
```

Figure 10-10 shows how our Cafeteria Portal home page looks on a BlackBerry 7100. Figure 10-11 shows how it looks on a BlackBerry 6230.

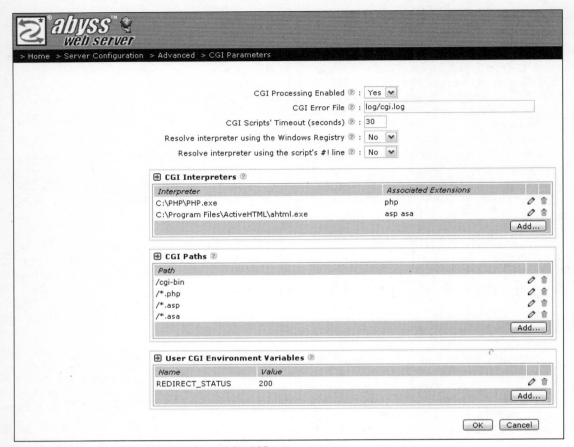

Figure 10-9: Abyss Web server configured for ASP support

Figure 10-10: Cafeteria Portal home page
on a BlackBerry 7100

**Figure 10-11: Cafeteria Portal
home page on a BlackBerry 6230**

You are probably wondering why the home page is also an ASP page. Technically speaking, the only page that needs to have any ASP code in it is the default page (which, in our case, is called `Default.asp`) and that ASP code is used to redirect the browser to the correct home page for each browser. For informational purposes only, we included a bit of ASP code that displays the browser type at the bottom of each home page.

If you scroll down to the bottom of the home page on the simulated BlackBerry, you will see a line that shows you that information. In a real portal page, you would only need to create regular HTML unless, of course, you wanted to use ASP in other ways. The Cafeteria home pages are all constructed using Microsoft FrontPage.

Feedback Form

While you are at the bottom of the Cafeteria home page, click the General Feedback link. This takes you to a simple feedback form that asks for your name, gives you a choice of a Yes or No answer using option buttons, and enables you to type in some comments. Again, no special attention was paid to the BlackBerry when this form was created. You can fill out the form and submit it if you like. If you want this form to be functional, you will need to change the submit e-mail address in the form. To do this, modify the form properties in FrontPage or open each page in Notepad and edit the S-Email-Address value to reflect your e-mail address.

For this demonstration, we are using two suggestion pages: one for the color BlackBerry devices and desktop browsers (see Figure 10-12), and one for the black-and-white BlackBerry devices (see Figure 10-13). In reality, you could just use one form because MDS does the transcoding necessary to strip out any unwanted HTML tags. For example, the color version of this form has a blue background and uses white text. If a black-and-white BlackBerry opened this page, MDS and the BlackBerry browser will ignore all color formatting and display black text on a white background.

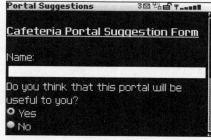

**Figure 10-12: Suggestions page for color
BlackBerry devices**

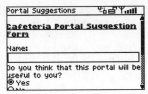

Figure 10-13: Suggestions page for monochrome BlackBerry devices

A feature of the BlackBerry 4.0 browser is the way in which it can handles Web forms. You can add extra parameters within your WML or HTML code that tell the BlackBerry browser to queue up your forms and send them later if you are out of coverage. The following table describes the extra parameters.

Parameter	Required	Description
x-rim-queue-id	Yes	Specifies the Offline Form Queue to which any GET or POST requests from form submissions on this page should go. The value may be any text string.
x-rim-next-target	No	Specifies the next page to load after sending any GET or POST requests resulting from this page to the Offline Form Queue. The value may be any valid URL.
x-rim-request-title	No	Specifies the label used to identify this request in the Queue view page. The value may be any text string. By default, the request is identified using the title of the page.
x-rim-request-id	No	Specifies whether the browser will generate a unique ID and add it as an HTTP header for every offline request resulting from this page. The value may be a Boolean True or False. By default, this value is True.
x-rim-request-date	No	Specifies whether the browser will generate a time stamp and add it as an HTTP header for every offline request resulting from this page. The value may be a Boolean True or False. By default, this value is True.

If you are writing in HTML or XHTML, these extra parameters can be inserted as hidden "input" elements.

In the Cafeteria Suggestion form, we added two extra parameters. The following tells the browser to create a new queue called "Cafeteria":

```
<input type="hidden" name="x-rim-queue-id" value="Cafeteria" />
```

The following tells the browser to name this form "Suggestion Form" when it is waiting in the queue.

```
<input type="hidden" name="x-rim-request-title" value="Suggestion Form" />
```

If you are writing in WML, these extra parameters can be inserted using `<postfield>` elements. For example:

```
<input type="hidden" name="x-rim-queue-id" value="Register" />
<input type="hidden" name="x-rim-request-title" value="Stock Monitor" />
<input type="hidden" name="x-rim-next-target" value="success.wml" />
```

Unfortunately, you cannot see these suggestion forms in action if you are using the Abyss Web server. It does not have any CGI scripts to handle form submissions. You can simulate how the offline submissions would work by running the MDS simulator and BlackBerry simulator. Next you can load the suggestion form page, switch back to the home screen by Ctrl+right-clicking (which would perform an Alt+Esc combination on a real BlackBerry), and turn off the radio. Then, switch back to the form, fill it out, and submit it. As shown in Figure 10-14, you will see that it is placed in a new queue called *Cafeteria*.

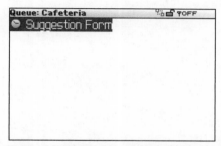

Figure 10-14: BlackBerry Offline Queue

If you then switch back to the home screen and turn the radio back on, when you switch back to the form, it will submit. If you open the form, you should see the thank-you page indicating that the form was successfully submitted.

As shown in Figure 10-15, you can go back to the Offline Queues at any time by opening the browser, clicking the scroll wheel, and choosing Offline Queues.

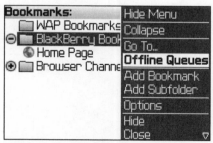

Figure 10-15: Opening Offline Queues

Summary

The BlackBerry browser is a versatile piece of software that makes the development of mobile Web pages very easy and enables the BlackBerry user to browse non-mobile Web sites from their BlackBerry device. In conjunction with MDS, the user experience is very similar to the desktop browser because MDS does a significant amount of transcoding of images and HTML.

As a developer, because of this great synergy between MDS and the BlackBerry browser, your learning curve is not as large as you once thought.

In Chapter 11, we will discuss how to create a BlackBerry channel. These channels provide a way for the user to have access to Web content tied to an icon on the BlackBerry home screen.

The BlackBerry Channel

The idea of a BlackBerry channel is to provide the user with a way to get to an intranet Web site with one click. You create an intranet portal Web site that is designed to look and work well with the BlackBerry. You then push the channel out to your BlackBerry population through MDS. MDS has built-in functionality for channels. As long as you format your MDS command correctly, MDS will know how to process the request and send out the channel.

The channel consists of two icons (a read icon and an unread icon) and a Web page. The Web page that you create must be BlackBerry-friendly — you can use regular HTML, as long as you pay attention to the screen real estate of the handheld unit and design the page accordingly. Your other option is to design the page using Wireless Markup Language (WML).

When the channel arrives on the BlackBerry, it will show up as a new icon on the home screen (also known as the *ribbon*). If the Web page is new (or has been changed), the unread icon will be shown. After the user has seen the Web page, the read icon is displayed.

The channel has an immediate benefit to the user. The Web page that you choose to push along with the channel can either contain content of its own or direct the browser to your intranet portal page.

The channel can be used as a means to deliver company news or emergency information. You can also use it simply as a springboard to your intranet portal, which may contain the same information, or it could provide the user with access to many internal resources that would have WML or carefully designed HTML interfaces.

Using the BlackBerry Channel

A BlackBerry channel can be used in two ways. You can create a channel that is self-contained. When the user clicks the channel icon, the BlackBerry shows the user a cached Web page that resides on the handheld device. This is useful because it does not require that the user have a wireless signal to open the Web page. If you designed this channel Web page using WML (as discussed in Chapter 10), WML pages can display as multiple pages (or cards in a deck). This means that a complete "mini-site" could reside on the handheld device.

This scenario is useful when you need to have one or many pages of useful information on every user's handheld device. An example of this could be company news, emergency contact information, cafeteria menu, and so on. The channel content is refreshed daily, hourly, or at any time interval, depending on the nature of that information. If the page is WML and contains many cards, the user will have access (no matter if the user has a signal or not) to the full or partial handheld portal. As you make changes to the WML and push out the changes, users will have access to the changes, as long as at some point their handheld devices were in an area with coverage to receive them.

The second use of a channel is as a springboard to live content on your intranet. When the user clicks the icon, the page that is loaded could simply redirect the browser to the live intranet Web site. Of course, this would require that the handheld unit have a wireless signal. Your BlackBerry should have a wireless signal most of the time, but there are instances where tunnels are not wired for cellular coverage, or there are dead zones in your carrier's coverage that would prevent the user from browsing to this content successfully.

As mentioned earlier, with the BlackBerry platform, you can use regular HTML to construct your Web pages because the MDS service strips out much of the formatting when it passes the pages on to the BlackBerry. Even though you can use HTML, you must test it thoroughly using the BlackBerry Simulator to see how it renders on the handheld device.

Creating Your First Channel

As a recap, a BlackBerry channel consists of two icons and a Web page. The two icons are for visually displaying when the channel is new or has been updated (unread), or when the content has been viewed (read). The Web page is stored in the BlackBerry browser's cache. Because it is stored in cache, it does not need a wireless network to function. This enables you to create content that your users can interact with at any time. Of course, your channel may have some links to Web sites and, when your users click these links, they must have a wireless signal to allow those pages to load.

In keeping with the Cafeteria Portal theme we are going to create two channels. One will simply be a springboard to launch the browser and connect to the Cafeteria Portal itself, and the second will be a more functional channel that will display the menu of the day.

The Channel Web Page

When you push a channel out to the BlackBerry devices, you must include a Web page. For this first channel, we will use WML. We will do this because the Web page will be much smaller and we only need to display a message and launch the cafeteria portal. Following is the WML Web page that we will push out to the handheld devices:

```
<?xml version="1.0"?>
<!DOCTYPE wml PUBLIC "-//WAPFORUM//DTD WML 1.2//EN"
"http://www.wapforum.org/DTD/wml_1.1.xml">

<wml>
<card ontimer="http://localhost/">
<timer value="10"/>
<p><big>Launching Portal</big></p>
```

```
<p><small>Please wait ...</small></p>
</card>
</wml>
```

Start at the top and work through the file. Consider the following:

```
<wml>
<card ontimer="http://localhost/">
<timer value="10"/>
```

The first line of this code shows the start of the WML file, which is indicated by the `<wml>` tag. The next two lines work together to wait a short period and then launch the browser to the URL `http://localhost/`. Remember that in this example we are using a Web server that is located on our PC, and so we use the URL of `localhost`. In a real-world example, you would want to use the real URL of your handheld portal site.

The statement `<card ontimer="http://localhost/">` tells the browser that a timer will be specified and when the timer runs out, to launch `http://localhost/`. The statement `<timer value="10"/>` tells the browser that the timer must run for one second. (The timer value is in increments of tenths of a second, so a value of 10 represents 1 second.)

Now, consider the next section of code:

```
<p><big>Launching Portal</big></p>
<p><small>Please wait ...</small></p>
</card>
</wml>
```

The first line in this section tells the browser to display the text "Launching Portal" using the "big" font. The second line tells the browser to display the text "Please wait ..." using the "small" font. Finally, the WML file is completed by closing the card and closing the `</wml>` tag. (See Chapter 10 for a discussion on the relationship of cards and WML.)

This page is rather simple in design, and there is no need to create a color and monochrome version. You can use this same page on either handheld type.

The Channel Icons

The purpose of the icons is to show a visual representation of the channel and to indicate whether the content is read or unread. For this first channel, the content that we are pushing to the handheld device is never going to change. This means that we can use the same picture for both read and unread icons. However, for the purpose of demonstration, we will use both icons.

When you create your read and unread icons, you may want to create a separate set of monochrome and color icons to take advantage of the two kinds of handheld devices, or you may simply want to create one set. If you have monochrome handheld devices in your environment, you may want to create a set of monochrome icons and push them to both monochrome and color handheld units. Depending on your artistic skills, you may be able to create a very appealing monochrome icon that looks good on both handheld devices.

One thing to keep in mind when using monochrome icons is that when certain BlackBerry themes are used (for example, the Vodacom Theme), the black in the icon can change color. For example, if you have a Vodacom BlackBerry that uses the Vodacom theme, a monochrome icon will become red and white. A BlackBerry theme is similar to a Windows theme. It modifies the device so that certain colors, icons, and fonts are used for all of the device menus and screen layouts. At this time, only RIM and cellular carriers can create themes, but hopefully this will change to allow for user-created themes.

Remember the way in which the handheld device deals with icons. When the icons are sent to the handheld unit, they are held in the browser cache. If you decide to change the look of the icons, you must rename the source PNG files. This is because of the way in which the cache is used. If you change the look of the source icons, but keep the file names the same, the handheld device will keep displaying the old icons that it has stored in cache (because it will assume that the new ones that are pushed out have not changed). By changing the file names of the source icons, you force the handheld unit to display your new changes. While you experiment with the icons, you can always delete the channels and then the browser cache by going into the browser options menu. This will effectively wipe out the channel and the cached icons.

If you only have color handheld devices in your organization, then you can simply create one set of color icons. If you want to create one set of monochrome icons and then one set of color icons, you will somehow need to know which users have what handheld devices so that you push the correct channel to them. If a user receives a new handheld unit, you must keep track of whether the user has switched from a monochrome to a color handheld device and push out the correct channel.

If you are in a pre-4.0 BES environment, then make use of a third-party asset trail database of some kind that you can query to find out what kind of handheld device a particular user is using before pushing out the correct icons for the channel. If you are already using BES 4.0, you can query the existing configuration database for the handheld properties.

For this discussion, we will create two versions of the channel: one for monochrome handheld devices and one for color handheld devices.

For color handheld devices, the icon dimensions are 32 × 32 pixels. For monochrome handheld devices, the icon dimensions are 28 × 28 pixels. One disappointing aspect of creating a channel icon is that it does not animate like the regular BlackBerry icons. If you scroll over a BlackBerry icon, you will see that it enlarges when it is in focus and shrinks when it is not in focus. Perhaps RIM will add a way for us to do this in the future (which would be great), but for now, your only option is a static icon.

When you design your icons, use a graphics program such as Paint Shop Pro or something similar. The reason for this is that these programs enable you to save images in many different formats, including Portable Network Graphics (PNG). PNG graphics files are what you need to push out to your handheld units. Another reason to use these kinds of graphics programs is that they enable you to make part of the image transparent. While you do not have to make use of transparency, it provides a much better look to the icons. All of the regular BlackBerry icons have parts that are transparent. When you navigate the home screen, a blue circle acts as your cursor. When you scroll onto an icon, you can see the icon and the circle. If your icon has no transparency, you will not be able the see the circle, and this just does not look appealing on the screen. In addition, the user could "lose" the cursor, because it hides behind the icon.

The idea of image transparency is using a graphics program to assign one color in the image to be used as the transparent color. When that image is displayed in a browser, the browser knows that any pixels that use that particular color must become transparent. The transparent color information is stored with the file when you save it. Remember that the browser is simply going to make any pixel it sees with that color transparent. If you want the background of your icon to be transparent, then you should not use the same color as the background color and in your icon image. For example, if you choose a particular shade of green as your transparent color, do not use that shade of green in your icon. If you do, your icon may become partly transparent. The best trick is to choose a color that you are not using in the actual icon itself and paint the background using that color, and then choose that color as the transparent color.

When you uncompressed the 589539 ch10 code.zip file earlier and copied all of the files to the root of your Web server, you would have noticed that we have included all of the Web pages and icons for the channel. We have also included a PHP script that will allow you to push this channel to your BlackBerry or the simulated BlackBerry. If you have not yet downloaded this ZIP file, visit the Wrox Web site to download it. Refer to the next section for instructions on how to install a local Web server if you do not already have one to use.

Preparing to Push the Channel

During this discussion, we will use the MDS simulator and the BlackBerry simulator to demonstrate how the MDS channel push works. Because we use Active Server Pages (ASP) and PHP during this demonstration, you will need to enable these two technologies on your Web server. You can install Microsoft IIS on your PC (as long as you are using Windows XP Professional) and use it as your local Web server. Or, if you do not have this capability, you can install a third-party Web server along with third-party PHP and ASP interpreters.

A free third-party Web server that installs on any version of Windows is called the Abyss Web Server X1 from www.aprelium.com. It is very fast and efficient and enables you to add third-party Common Gateway Interface (CGI) interpreters. You can download a third-party ASP interpreter from seliSoft at www.selisoft.com and a third-party PHP interpreter from www.php.net/downloads.php.

After you have installed the Abyss Web server, execute the following instructions to install PHP and ASP support.

Add PHP Support to the Abyss Web Server

After you have downloaded the PHP support from www.php.net/downloads.php, start the installation. If the installation process asks what kind of Web server you are using, choose None (or other Web server), I will configure the Web server manually.

When the installation is complete, open the Abyss Web console and click Server Configuration. On the Server Configuration screen, click the Advanced button, and then CGI Parameters. Next, click Add in the CGI Interpreters table. In the Interpreter Path field, enter the path of the php.exe executable file (for example, c:\program files\php\php.exe). Enter **php** in the Associated extensions field, set Automatic Update of the CGI Paths (if available) to Yes, and click OK. Click Add in the User CGI Environment Variables table. Enter **REDIRECT_STATUS** (with no leading or trailing spaces) in the Name field and **200** in the Value field. Click OK.

Add ASP Support to the Abyss Web Server

After you have downloaded ActiveHTML from www.selisoft.com and installed it, you must configure the Abyss Web server to use it.

Open the Abyss Web Server's console, select Server Configuration, click Advanced, and select CGI Parameters. Ensure that CGI Processing Enabled is set to Yes. Next, click Add in the CGI Interpreters table. In the Interpreter Path field, enter the path of the `ahtml.exe` executable file (for example, `c:\program files\selisoft\activehtml\ahtml.exe`). Enter **asp asa** in the Associated extensions field, set Automatic Update of the CGI Paths (if available) to Yes. Click OK.

Figure 11-1 shows the Abyss Web server configuration after PHP and ASP support have been added. Remember to restart the Web server to activate the changes.

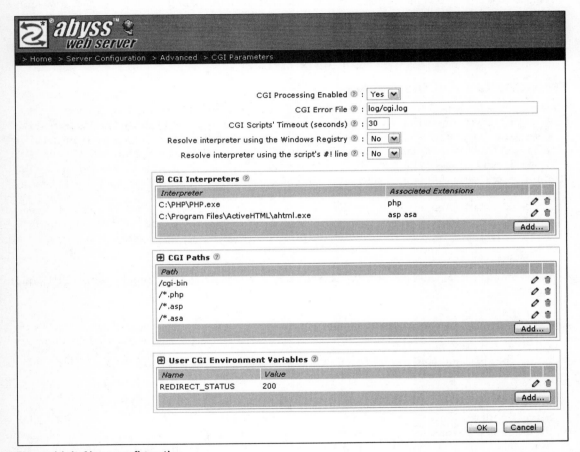

Figure 11-1: Abyss configuration

How Does the Push Work?

When you want to push something to a BlackBerry handheld device, send a POST command to the MDS service and supply certain information about the pushed item, including the user's e-mail address. The MDS service keeps a table that links the e-mail address to the handheld PIN. You can push content to handhelds using only an MDS service that has been set as a push service.

To issue an MDS PUSH command, you need at least the following information:

- ❑ BES DNS name or BES IP address

- ❑ BES MDS push port (normally 8080 for the Simulator, Novell GroupWise, and a Lotus Domino BES, and 8300 for an Exchange BES)

- ❑ Recipient's e-mail address

- ❑ URL to be pushed

- ❑ Push Type (`Browser-Channel`, `Browser-Message`, `Browser-Content`, `Browser-Channel-Delete`)

- ❑ Push Title (name of the channel or subject of the browser message)

- ❑ Unread icon URL

- ❑ Read icon URL

In our example, we use a simple HTML form to gather this information, and then we post it to a PHP script that takes the data and sends it to the MDS service using the correct commands. The data is passed to MDS using the following POST commands:

- ❑ `X-Rim-Push-Type`: Specifies how the content is pushed to the browser (`Browser-Message`, `Browser-Content`, `Browser-Channel`, `Browser-Channel-Delete`).

- ❑ `X-RIM-Push-Title`: Name of the channel or subject of the browser-message.

- ❑ `X-RIM-Push-UnRead-Icon-URL`: The URL to the unread channel icon.

- ❑ `X-RIM-Push-Read-Icon-URL`: The URL to the read channel icon.

- ❑ `Content-Location`: Location of the content to be pushed. This is one of the few parameters that does not begin with *X-Rim*.

After the MDS service has accepted the PUSH command and has successfully submitted it, it will send back a confirmation page. If an error occurs during the push, the MDS service will send back an error page. In our example, the PHP script captures this MDS output and displays it on the screen.

Pre-4.0 MDS does what is called an *unreliable push*. When you submit your push request, the MDS service acknowledges that it has received it, but it does not acknowledge when it has been successfully delivered to the handheld device. Sometimes, the devices can be out of coverage for longer than the default 10-minute timeout period and that device will not receive the push.

With MDS 4.0, you can include extra parameters in the PUSH command that instructs MDS to confirm when the content has successfully reached its destination. In addition to this method of reliable content pushing, you can specify extra commands that determine how the device acts when it receives a push. Following are these extra commands:

- ❑ `X-Rim-Push-Ribbon-Position`: Specifies in what position to place the channel on the BlackBerry *ribbon* (also known as the *Home Screen*). Note that because of a bug, this parameter does not currently work. RIM plans to correct this in a future software fix.

- ❑ `X-Rim-Push-Priority`: Specifies if the user receives notification on the device if content is received.

❑   `High`: A dialog box is displayed on the screen notifying the user that the content has been received, or that an update has been received. The Browser notification method is also triggered. This notification method can be set by the user in the Profile application. Last, an unread icon is placed on the Ribbon.

❑   `Medium`: The Browser notification method is triggered and an unread icon is placed on the ribbon.

❑   `Low`: Same as Medium.

❑   `None`: An unread icon is placed on the Ribbon.

❑ `X-Rim-Push-Reliability`: Specifies the reliability of the push.

❑   `Transport`: A message is sent from the device once the content has arrived. This will work on any handheld unit.

❑   `Application`: A message is sent from the device when the content has arrived at the destination application on the device. This only works on handheld devices running the 4.0 code.

❑   `Application-Preferred`: If the device is running 4.0 code, it will send an application response. If the device is running pre-4.0 code, it will send a `Transport` response.

❑ `X-Rim-Push-NotifyURL`: Used in conjunction with the `X-Rim-Push-Reliability` command, the responses will be sent to the URL specified. A script of some kind (PHP, ASP, and so on) at that URL must capture the response.

❑ `X-Rim-Push-Deliver-Before`: Specifies a date by which the content must be delivered. The date is in standard HTTP format.

❑ `X-Rim-Push-ID`: Specifies a unique value that you use to identify the push. This value can be used later to query the MDS service on the status of the push.

❑ `X-Rim-Transcode-Content`: Used to override the standard transcoding rules of MDS.

❑   `*/*`: MDS transcodes all content.

❑   `None`: MDS does not transcode any content.

❑   `<list of MIME types>` (for example, image/jpeg, image/png): MDS only transcodes these MIME types.

❑ `Content-Type`: Defines which MIME types can be included with the content push (for example, image/jpeg, image/png).

❑ `Cache-Control`: Overrides the normal handheld browser cache behavior for the pushed content.

❑   `No-cache`: No part of the pushed content is cached.

❑   `Max-age`: Maximum age (in seconds) before the cached content expires.

❑   `Must-revalidate`: The content is always loaded from the Web server and never loaded from cache.

Pushing the Channel

To send our Cafeteria channel to our simulated BlackBerry, we must start up the MDS simulator and the BlackBerry simulator. By default, the MDS simulator has a small table of e-mail addresses linked to a small list of simulator PIN numbers. If you have not made any changes to the MDS simulator or device simulator configuration files, the device simulator should start up as a BlackBerry 7290 with the PIN of 2100000a. The MDS simulator links this PIN number to the e-mail address `simulator@pushme.com`.

If you have changed any configurations, you may need to edit the MDS configuration and add (or edit) the PIN numbers and e-mail address. To do this, edit the file `rimpublic.property`, which you will find in `c:\program files\research in motion\blackberry jde 4.0\mds\config`. Scroll down to the `# [Simulator]` heading and modify the list of PIN numbers and e-mail addresses.

For this demonstration, we will assume that you are using the BlackBerry 7290 simulator with a PIN number of 2100000a.

Start up the MDS simulator and the BlackBerry simulator. Next, open your favorite Web browser and open the page called `push.htm` on your Web server. If you are using a local Web server (such as Microsoft IIS or Abyss) the URL will be `http://localhost/push.htm`. If you are using an external Web server, the URL will be `http://<Web server name>/push.htm`.

You should see the push page shown in Figure 11-2.

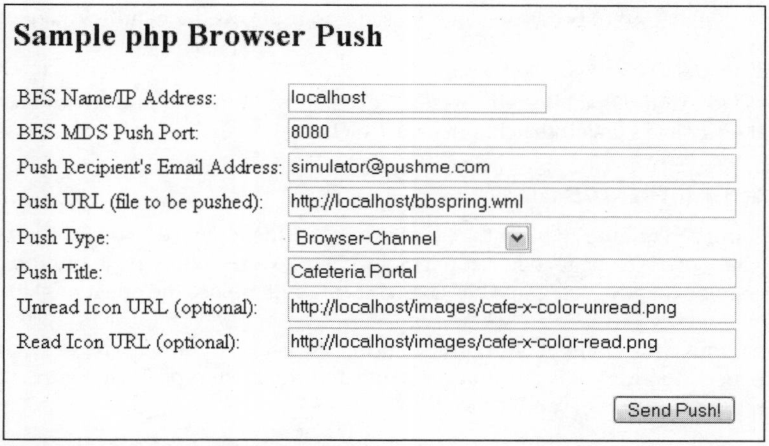

Sample php Browser Push

BES Name/IP Address:	localhost
BES MDS Push Port:	8080
Push Recipient's Email Address:	simulator@pushme.com
Push URL (file to be pushed):	http://localhost/bbspring.wml
Push Type:	Browser-Channel
Push Title:	Cafeteria Portal
Unread Icon URL (optional):	http://localhost/images/cafe-x-color-unread.png
Read Icon URL (optional):	http://localhost/images/cafe-x-color-read.png

Send Push!

Figure 11-2: PHP push Web page

Most of the fields have already been filled in, so type in these values into the following fields:

- ❑ BES Name/IP Address: **localhost** (or you can use **127.0.0.1**)
- ❑ BES MDS Push Port: **8080** (or **8300** for Exchange)
- ❑ Push Recipient's Email Address: **simulator@pushme.com**
- ❑ Push Title: **Cafeteria Portal**

Click the Send Push button. After a few seconds, the results screen will appear, indicating that the push was successful. If you switch to your BlackBerry simulator, you will see a new icon on the ribbon. Figure 11-3 shows what it should look like.

Figure 11-3: Channel pushed

Using the Channel

Before you click the new channel icon, notice that it has a red background. This is because you have not viewed it yet, and so the handheld device is displaying the unread icon. Click the icon. The BlackBerry browser launches and opens the BBSpring.wml Web page. This page displays a message for one second and then attempts to open the Cafeteria Portal page at http://localhost/. As in Chapter 10, this first page determines what kind of browser is being used and redirects the handheld device to the appropriate page.

Press ESC twice to exit the BlackBerry browser. Now you will notice that the icon has a blue background, indicating that you have viewed the channel since a change occurred.

Other Channel Functions

After you have pushed out the channel to one or many handheld devices, you do have options. If you want to update the channel's Web page, you can make the changes and simply push out the channel again. The device will recognize that the Web page has changed and will replace the one it has stored in cache.

If you want to remove the channel, you can perform a Browser-Channel-Delete command. When you do this, all you need to provide is the channel name and the push type of Browser-Channel-Delete. You do not need to specify the icon URLs.

More on Reliable Push and Other Push Functions

Now that you have successfully pushed out your first channel, you can experiment with more advanced push functions. If you load the page http://localhost/push-adv.php, you will see that there are some extra fields in the form, as shown in Figure 11-4. These fields allow you to send extra MDS PUSH commands to control how the channel (and other push content) is pushed to the device, and how the device will react upon receiving the content.

Additional Options

Ribbon Icon Position (optional):	☐ 1
Push Description (optional):	☐ Test Push
Push Priority (optional):	☐ Low ▾
Push Reliability (optional):	☐ Transport ▾
Push Notification URL (optional):	☐ http://localhost/phpAccept.php
Deliver Before (optional):	☐ 2005-04-21T21:28Z
Push ID (optional):	☐ 27851@yourServer.com
Transcode Content (optional):	☐ */*
Content Type(s)(optional):	☐
Cache Control (optional):	☐ no-cache ▾
	Max Age Value (seconds):

[Send Push!]

View push notifications here.

Figure 11-4: The push-adv.php page

Push Priority

If you push content to the BlackBerry, you can change the behavior of the device when it receives it. As per the previous table, if you set the X-Rim-Priority to High, the handheld unit will trigger the browser notification, add the icon to the ribbon, and pop up a dialog box indicating that the content has been added or updated. The browser notification trigger is set by the user in the Profile application. By default, when the BlackBerry is out of its holster, it will not vibrate or tone. To see the effect of a high-priority push in the simulator, you must modify the "Out of Holster" behavior of the BlackBerry by editing the profile, as shown in Figures 11-5 and 11-6.

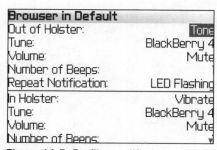

Figure 11-5: Profile modified

Figure 11-6: Result of high-priority push

Reliable Push

When you want to send content to the handheld device using a reliable push, you must add at least two extra commands. The first is X-Rim-Push-Reliability. This instructs MDS on how to push the content. As per the earlier table, if you sent X-Rim-Push-Reliability as Transport, the handheld device will send back an acknowledgment when the content has arrived. Transport-level reliability works if you have a 4.0 BES or later, and any version of Handheld Code. If it is sent as Application, then the application to which you sent the content would send back the acknowledgment once it has been received. Finally, if it is sent as Application-Preferred, then the handheld will send back a transport-level acknowledgment if it is not running at least version 4.0 of the Handheld Code, and an application-level acknowledgment if it is running version 4.0 Handheld Code or later.

The second command is X-Rim-Notify-URL. If you do not send this command, then MDS has no way of notifying you once it receives the acknowledgment from the handheld device. This URL typically is a PHP or ASP page that can accept input from a Web POST.

In our example, if you specify the URL as http://localhost/phpAccept.php, the PHP script will save the reliable push results to the file called notifications.txt. If you open that file, you will see something similar to this:

```
Date: Wed, 30 Mar 2005 15:56:14 -0500; X-RIM-Push-ID: 1ba0f36:102f53f1559:-
7ffe76188; X-RIM-Push-Status: 200
```

Last, you can send a unique ID along with each content push, so that you can later query MDS on the status of that particular push request. This command is X-Rim-Push-ID. This last command is not required.

The Today's Menu Channel

The previous example of a channel was a channel that launched into a live Web site, requiring that the user have a wireless signal to browse the Web site. A BlackBerry channel does not have to be this simple, and the Web page that you push out with the channel can be self-contained if it is written in WML. As discussed in Chapter 10, a WML page is similar to a deck of cards. Each deck can contain one or many cards. Each card is seen as a new page to the user. With this next channel, we will demonstrate that you can create and push out a self-contained channel that the user can interact with without the need for a wireless signal.

Since we will use the same icons for this channel, to avoid confusion we should delete the current channel from your BlackBerry simulator. To do this, load the push page again and enter the relevant information for the previous channel. This time, however, change the push type to `Channel-Delete`. After the channel has been deleted, you can proceed.

Preparing the Channel

The idea for this channel is a daily menu. The channel would allow the user to read the cafeteria menu for that day, including the specials. The channel would be prepared and pushed out in the morning so that the employees could have time to read it before lunch. The user would be able to see that the new daily menu has been pushed out because the channel icon would change back to the unread icon (in our case, the icon with the red background).

Although our example is that of a daily menu channel, there are unlimited uses for a channel that can be viewed even when the user is out of wireless coverage. A good example is an emergency contact information channel. This is particularly useful if something happens that brings down the cellular network or something happens in the building that houses your BES. An emergency contact or emergency information channel would already be pre-loaded onto the handheld devices and the users would have access to this information.

Another example is internal company news. From an internal marketing standpoint, it may be advantageous to have a channel that shows the latest company news (such as what deals have recently been won by the company, has the company been featured in any news stories, and so on). This can prove to be a great marketing tool.

The WML page we prepared for this example contains four cards. The first card is the main menu, while the remaining three cards show menu choices. There is nothing stopping you from including a feedback form as an extra card. When you fill out a form on a BlackBerry and you are out of coverage, the form submission is held in a queue until the user returns to an area of wireless coverage. Our new channel page is called tmenu.wml. Figure 11-7 shows what it looks like.

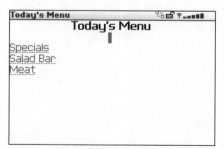

Figure 11-7: Daily menu WML

To push this channel to your simulated BlackBerry, return to the push page and enter the information as follows:

❑ BES Name/IP Address: **localhost** (or you can use **127.0.0.1**)

❑ BES MDS Push Port: **8080** (or **8300** for Exchange)

❏ Push Recipient's Email Address: **simulator@pushme.com**

❏ Push Title: **Cafeteria Menu for Today**

Change the Push URL field to `http://localhost/tmenu.wml` and click the Send Push! button.

When MDS confirms that the channel has been pushed, look at your BlackBerry simulator. Ensure that the channel is there and then close the MDS simulator. Closing the MDS simulator will prevent the BlackBerry from communicating through MDS. You can turn off the radio to simulate an out-of-coverage situation.

Click the channel. You will notice that it loads normally. You can browse the channel and you will be able to view the different menu pages. As you can see, the channel is fully self-contained.

Summary

This chapter discussed how to create and push out a BlackBerry channel. We have learned that we can create a simple channel that calls an external Web server, or we can create a channel that is completely self-contained and enables the user to navigate the channel without the need for a wireless signal.

While we used the idea of a Menu of the Day channel, the possibilities for these kinds of channels are endless and can provide a significant benefit to any organization.

In Chapter 12, we discuss the Web Message and Browser Cache MDS push commands, which will enable you to send a Web page directly to the device's inbox or to the browser cache.

The BlackBerry Web
Message and Cache Content

Chapter 11 discusses how to create a BlackBerry channel, which you can push out to your handheld devices. The BlackBerry Web message is another way in which you can push content out to your handhelds. This time, the content shows up as an item in the user's handheld inbox. This kind of push technique can prove very useful for content that you want users to see immediately as a new message, as opposed to users paying attention to a changing channel icon on the handheld ribbon.

A Web message (or browser message) does not reside in the browser cache like a channel does, but rather resides in the handheld inbox as a message. This means that it cannot be deleted unless the user chooses to delete it. If the user clears the browser cache, the Web message is not removed.

The BlackBerry Cache Content is a way to push a Web page to the handheld browser cache so that when the user requests that page (either directly or through a link in an existing Web page, channel, or Web message), it will be displayed without the need for wireless coverage.

Creating a BlackBerry-Friendly
Web Message

Because a Web message is a Web page that is pushed to the handheld inbox, the way in which you construct this Web page should follow the same rules as we have discussed in previous chapters. If you use HTML, you should remember that you are working with a mobile device that has limited screen real estate. If you use WML, you will be able to create a mini-site that the user can browse without the need for wireless coverage.

So far, we have used our imaginary Cafeteria Portal as a vehicle to show the different MDS concepts. This chapter is no exception. Our Cafeteria Portal already has a menu listing and we have already pushed the menu to the handheld devices as a channel, but let us pretend that we want to push this content to the handheld units as a Web message. The users in our imaginary company really need to see the menu for each day of the week so that they can plan accordingly. We have created a WML page to push out called `tmenu-mess.wml`. It is a copy of the `tmenu.eml` page, so there is no difference between the channel page you pushed out in Chapter 11 and this Web message.

When you push out a Web message, there are no associated icons to push. This is because the Web page will appear in the handheld inbox as a new Web message. The BlackBerry already has icons to show a new and read Web message and it will use them appropriately.

If you have not already done so, visit the Wrox Web site and download the file called `589539 ch10 code.zip`. Unzip the contents of this file to the root of your Web server. Refer to Chapter 10 or Chapter 11 for further instructions on how to install a local Web server if you do not already have one.

To push out our Web message, launch the MDS simulator and the BlackBerry simulator. Next, open your favorite Web browser and open the page called `push.htm` on your Web server. If you are using a local Web server (such as Microsoft IIS or Abyss), the URL is `http://localhost/push.htm`. If you are using an external Web server, the URL is `http://<Web server name>/push.htm`.

Fill out the fields as shown here and in Figure 12-1. Remember, you do not need to supply the read and unread icon URLs because they are not used. If you accidentally specify URLs in these fields, MDS will simply ignore them.

❑   BES Name/IP Address: **localhost** (or **127.0.0.1**)

❑   BES MDS Push Port: **8080** (or **8300** for Exchange)

❑   Push Recipient's Email Address: **simulator@pushme.com** (or the relevant user email address)

❑   Push URL (file to be pushed): **http://localhost/tmenu-mess.wml**

❑   Push Type: **Browser-Message**

❑   Push Title: **Menu for Today**

Sample php Browser Push	
BES Name/IP Address:	localhost
BES MDS Push Port:	8080
Push Recipient's Email Address:	simulator@pushme.com
Push URL (file to be pushed):	http://localhost/tmenu-mess.wml
Push Type:	Browser-Message
Push Title:	Menu for Today
Unread Icon URL (optional):	
Read Icon URL (optional):	
	Send Push!

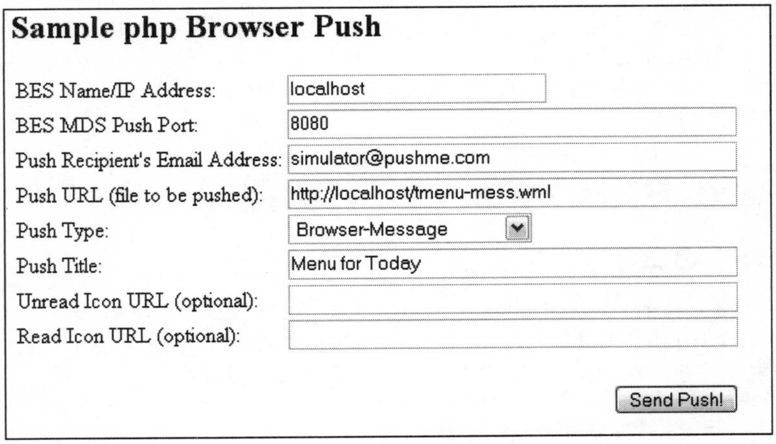

Figure 12-1: Push page

Click the Send Push! button and switch to your device simulator. You will see a new icon appear above the battery indicator on the home screen indicating that a new Web message has been received and is unopened. Figure 12-2 shows this icon.

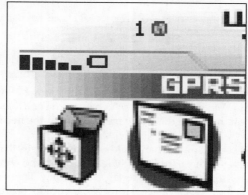

Figure 12-2: Unread Web message

Open the Messages application and you will see a new Web message. Instead of the icon of an envelope, you will see a tiny globe icon. This is the unread Web message icon. If you have users with monochrome handheld devices, the icons will be a little different. The monochrome icon that indicates a new unread Web message on the handheld ribbon is a tiny W. A slightly larger W icon is used to indicate a Web message in the Messages application. Figures 12-3 through 12-5 show the icons.

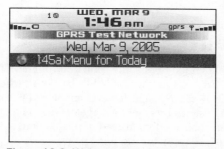

Figure 12-3: Web message in inbox (color)

Figure 12-4: Unread Web message (monochrome)

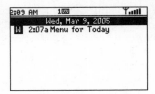

**Figure 12-5: Web message
in inbox (monochrome)**

Now click the new Web message and you will notice that the menu has a new item called Open Page. Click this menu item and the Web message opens. As before, the entire page is in the handheld's local browser cache, so the user can navigate the entire menu without the need for wireless coverage.

While the Web page itself is cached on the handheld unit, any images that you include in the Web page are not. If you do include images, those images will load up over the wireless network if the handheld device is in coverage. Otherwise, the image's `alt` tag will display. There is certainly nothing wrong with using images to enhance the user's experience. However, if you do, ensure that you use descriptive `alt` tags. For example, this line of code embeds an image into a WML page:

```
<img src="pic.png" alt="My Picture" />
```

If the user were out of wireless coverage when opening this page, the user would see [My Picture] in place of the actual image. After the image has been viewed at least once while the handheld device is in wireless coverage, it will reside in the browser cache. If the user then views the same page while out of coverage, the image will be retrieved from cache.

Browser-Content

As mentioned earlier, the final method of pushing content through MDS is the `Browser-Content` push. This method pushes content directly into the handheld's browser cache so that when that particular page is referenced in a channel or browser-message, it can be displayed without requesting it on the wireless network. This is useful if you design a browser message or channel that has mostly static content, but includes one or two pages that change often. Instead of pushing out a new browser-message or channel each time, you can simply push out only the modified pages.

In the case of a channel, the channel icon remains in the read state, indicating no change to the channel. However, if the user is browsing that channel and the user requests a certain page that you recently modified and pushed through the `Browser-Content` method, the new data will be displayed.

Preparing Browser-Content

`Browser-Content` follows the same rules that are applied to creating any content for the BlackBerry. To show how this works, we have created a new version of the Cafeteria Portal that has a new link on it. That link displays the chef on duty for that day. We have a separate page that displays the chef of the day and we will push that page into the browser cache through a `Browser-Content` push.

Pushing Out the New Channel

In the real world, you would make changes to your existing channel and push it out again. In our example, we are using a new WML file name. If we push out a channel with the same name but with a different URL, the device will see this as a new channel. To avoid this in our example, you will first need to remove the current channel from the device. To do this, load the page `http://localhost/push.htm`. Fill in the fields as follows:

- ❑ BES Name/IP Address: **localhost** (or you can use **127.0.0.1**)
- ❑ BES MDS Push Port: **8080** (or **8300** for Exchange)
- ❑ Push Recipient's Email Address: **simulator@pushme.com**
- ❑ Push Title: **Cafeteria Portal**

Change the Push-Type to `Browser-Channel-Delete`. Now, click on the Send Push! button. Make sure that the channel has been removed from your handheld device.

After it has been removed, reload the page `http://localhost/push.htm` (or simply click Back in the browser). Enter the information as follows:

- ❑ BES Name/IP Address: **localhost** (or you can use **127.0.0.1**)
- ❑ BES MDS Push Port: **8080** (or **8300** for Exchange)
- ❑ Push Recipient's Email Address: **simulator@pushme.com**
- ❑ Push Title: **Cafeteria Portal**
- ❑ Push URL: **http://localhost/tmenu-chef.wml**
- ❑ Push Type: **Browser-Channel**

Click the Send Push! button. You should see the channel on your simulator. Do not open it just yet. We need to push out the new page to the browser cache.

Pushing Out the Browser-Content

Reload the page `http://localhost/push.htm` (or simply click Back in the browser). Enter the information as follows:

- ❑ BES Name/IP Address: **localhost** (or you can use **127.0.0.1**)
- ❑ BES MDS Push Port: **8080** (or **8300** for Exchange)
- ❑ Push Recipient's Email Address: **simulator@pushme.com**
- ❑ Push Title: **Cafeteria Portal**
- ❑ Push URL: **http://localhost/chef.wml**
- ❑ Push Type: **Browser-Content**

Click the Send Push! button.

Testing the Browser-Content

Before we load the new channel, shut down the MDS Simulator by clicking the X icon in the upper right-hand corner of the MDS Simulator window. If you like, you can turn off the radio on the simulated BlackBerry.

Now that there is no way that the simulated BlackBerry can communicate with the Internet, load the channel. You will see a new item labeled Chef on duty today. Click that item and the page displaying that chef on duty will be displayed. Close the browser on the handheld device.

Now, start up the MDS Simulator again and turn on the radio on the BlackBerry. Edit the page `chef.wml` and change the name of the chef on duty. Save the file. Reload the page `http://localhost/push.htm` (or simply click Back in the browser). Enter the information as follows:

- ❏ BES Name/IP Address: **localhost** (or you can use **127.0.0.1**)
- ❏ BES MDS Push Port: **8080** (or **8300** for Exchange)
- ❏ Push Recipient's Email Address: **simulator@pushme.com**
- ❏ Push Title: **Cafeteria Portal**
- ❏ Push URL: **http://localhost/chef.wml**
- ❏ Push Type: **Browser-Content**

Click the Send Push! button.

Close the MDS Simulator again. Open the channel on the BlackBerry and click the link that shows the chef of the day. You will see that the page has changed. This proves that you can update only the pages you want on the handheld device without updating the whole channel. These pages can also be browsed while out of wireless coverage.

Summary

This chapter discussed how to send push content to the handheld devices through the `Browser-Message` (which is where you send a Web page to the inbox on the handheld unit) and the `Browser-Content` (where you update certain Web pages already in the browser cache or add new ones to the cache).

Both of these push methods have their own set of uses. Used separately, or in conjunction with each other, you can create very useful content for your mobile users, as well as keeping them up-to-date even when they are out of wireless coverage.

In Chapter 13, we discuss writing Java (J2ME) applications for the BlackBerry. We will also discuss how to distribute the applications once they have been written and tested.

13

Developing BlackBerry Java Applications

One of the most powerful features of the BlackBerry wireless device is that standalone BlackBerry applications can be developed using the Java 2 Micro Edition (J2ME) language.

Standalone applications differ from Web-based solutions in that they can be designed to work even when network coverage is unavailable. The primary benefit of creating a standalone application is that the developer has complete control of the look, feel, and provided functionality.

RIM provides all the tools and documentation required to create BlackBerry applications. Go to www.blackberry.com/developers/ to download the latest version of its Java Development Environment (JDE), both volumes of the *BlackBerry Application Developer Guide v4.0,* and all associated code samples. In addition, the *BlackBerry Developer Journal* (www.blackberrydeveloperjournal. com) provides a wealth of material in each issue that has proven to be of great benefit to developers.

Between the material covered in this book, RIM's tools and documentation, and the material published in the *BlackBerry Developer Journal*, you will have no difficulty ramping up in standard J2ME. After that, you can learn to unlock the full power of the BlackBerry through use of the BlackBerry Application Programming Interfaces (APIs).

Throughout this chapter, we will assume that you are already fluent in Java Standard Edition (J2SE) or Java Enterprise Edition (J2EE) programming. This chapter begins with an overview of basic concepts for programming with J2ME and progresses through an examination of how to prepare your first BlackBerry application.

J2ME Basics

Sun Microsystems has defined configurations and profiles within J2ME that include Connected Limited Device Configuration (CLDC), and Mobile Information Device Profile (MIDP). CLDC defines an underlying environment based on the physical capabilities of the device, whereas MIDP determines the core application functionality on the device.

J2ME is made up of a configuration (CLDC) and a profile (MIDP) that is specific to devices that share the following characteristics:

❑ Have a minimum of 128 KB to 512 KB of memory

❑ Have a 16-bit or 32-bit CPU

❑ Provide an implementation of a small Virtual Machine (VM)

❑ Are battery-operated with a low level of power consumption

❑ Have wireless network connectivity and restricted bandwidth

The main benefit of writing your wireless applications in standard J2ME is that it is compatible with all wireless devices that support the language. This means that your application will work on many current cellular telephones and wireless devices, including the BlackBerry.

The strength of this solution is also its weakness. Software written to be as generic as possible also tends to lack the character and style of an application written to take advantage of traits of the target platform. Choice of language should be determined by the needs of the end user. If the end user is expected to rely on a wide assortment of J2ME-enabled devices, then develop your application in standard J2ME and forget about taking advantage of the full power of the BlackBerry. If your target audience is BlackBerry users, then take full advantage of the platform by developing in J2ME with extensions through the BlackBerry APIs.

J2SE Versus J2ME Edition

J2ME was designed for use on restricted target platforms. As such, it does not include all of the functionality of J2SE, but does provide extensions for wireless coverage, telephone, Wireless Access Protocol (WAP) browsing, smaller screens, limited memory and storage, restricted power use, and more.

J2ME does not provide support for the following J2SE features:

❑ Floating-point operations

❑ Java Native Interface (JNI)

❑ Large user interface APIs such as Swing and Abstract Windows Toolkit (AWT)

❑ Thread groups

❑ Daemon threads

❑ Finalization

❑ Object serialization

❑ Reflection

❑ Remote Method Invocation (RMI)

J2ME provides three error classes, relies on Record Management System (RMS) persistent storage for file handling, and lacks a class loader. The latest release of the BlackBerry handheld software and JDE supports MIDP 2.0.

Now that we have attained some context for the J2ME, let us see how it integrates with BlackBerry. First, take a look at a valuable building tool, the APIs, and then we will discuss the BlackBerry JDE.

BlackBerry API Basics

RIM has developed a large, ever-expanding series of APIs that enable developers to take full advantage of features inherent on the BlackBerry wireless device. Many of the APIs can be used directly on a live device without difficulty, while others must be "signed" using keys provided by RIM, before they can be deployed. *Code signing* was implemented to prevent the spread of software that could cause difficulties for users and wireless network providers. Note that all applications (signed or otherwise) can be tested with the BlackBerry simulator without being signed. Signing is a straightforward process that is only required before live deployment.

The following table provides a brief look at many APIs that are supported with the latest release of the BlackBerry JDE.

Application Programming Interface (API)	Signed API?	Description
ApplicationMenuItem	Yes	Used to add custom menu items to BlackBerry applications such as the Address Book, Calendar, and messages.
Bluetooth	No	Provides serial communication support for Bluetooth connections.
Browser	Yes	Provides access to the BlackBerry Browser application to create and display HTML or WML pages and to set browser options.
Browser Field	Yes	Provides access to browser-component functionality to integrate in applications.
Browser Plugin	Yes	Defines the Plugin API used to add support for specific MIME-types to the BlackBerry Browser.
Collection	No	Defines basic functionality for data collections.
Compress	No	Provides components for compressing and decompressing data.
Crypto	Yes	Provides a collection of classes used to implement security for BlackBerry applications. The following tasks can be accomplished with the Crypto API: encrypt and decrypt data; digitally sign and verify data (secure the integrity of your data); authenticate data.

Table continued on following page

Application Programming Interface (API)	Signed API?	Description
Invoke	Yes	Used to invoke (run) BlackBerry applications such as phone, messages, MemoPad, tasks, and browser.
Internationalization	No	Used to support internationalization of applications.
IO	No	Provides a library of components to manage data I/O.
IO HTTP	Yes	Used to register with the BlackBerry Browser as the provider for one or more URLs.
IT Policy	No	Used to customize features that are common to all users on a given BES, such as password details, e-mail forwarding options, and browser settings. IT policies provide an efficient method to manage many different users simultaneously.
Lightweight Data Access Protocol (LDAP)	No	LDAP is a client-server protocol used for accessing directories. It enables you to access and browse through information on remote directory servers.
Low Memory	No	Maintains memory resources on the device when the device becomes over-taxed and runs low on memory.
Mail	Yes	Provides functionality for sending, receiving, and accessing e-mail and PIN messages.
Math	No	Provides a collection of fixed-point math routines and implementations of a stack for use in matrix multiplication.
Menu Item	No	Used to add menu items to custom applications.
MIME	No	Used for manipulating streams of MIME-encoded data.
Notification	Yes	Provides functionality to trigger event notifications through Consequence (tone, vibration, or LED) or NotificationsEngineListener (dialog boxes or interaction screens on the device).
Options	Yes	Used to add items to the handheld unit's Options screen.
PDAP	Yes	Provides interfaces that developers should implement to conform to the Personal Digital Assistant Profile (PDAP) specification on the BlackBerry wireless device (such as Address Book contacts and Calendar entries). Also see the PIM package from the MIDP specifications.

Application Programming Interface (API)	Signed API?	Description
Phone	Yes	Provides access to the BlackBerry Phone application to initiate calls, receive related notification events, and change phone options.
Phone Logs	Yes	Provides access to the historical logs of phone calls.
Plazmic Media Engine	No	Contains necessary functionality to download and view media on the BlackBerry.
Service Book	Yes	Used to store setup and configuration information on a BlackBerry and contains `ServiceRecords` that define information about BlackBerry connections. Service records are used to manage the handheld's CMIME (mail) and IPPP (Internet connections) and are also valuable for keeping track of network UIDs to manage more easily MDS connections.
Synchronization	Yes	Provides functionality required to synchronize data between the BlackBerry and the desktop software and contains classes and methods used to provide backup and restore functionality to applications.
System	Yes	Provides a variety of system-level functionality such as the following: interface classes that can be implemented to listen for various system-level events; base classes for application functionality and application management; base classes to describe persistent data objects and stores of such objects; system-level access to device services (such as the radio, phone, peripherals, available networks, event logger, memory manager, serial port, speaker, SIM card, and more).
User Interface	No	Provides functionality for constructing a user interface.
XML	No	Contains Simple API for XML (SAX) and WAP Binary XML (WBXML) parsers.

The BlackBerry Java Development Environment

You can develop standard J2ME applications by using any J2ME development environment you prefer. The best practice is to perform all of your development, editing, source control, testing, debugging, profiling, signing, and final compilation using the Java Development Environment (JDE) provided by RIM. The most critical JDE features, such as compilation, code signing, device simulation, e-mail server simulation, and Mobile Data Server (MDS) simulation, are not available with any other development environment.

Figure 13-1 shows the main JDE screen. Throughout this chapter, we will refer to menu options selected from this screen.

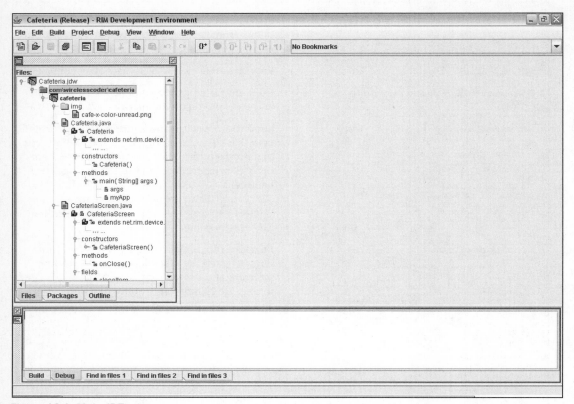

Figure 13-1: Main JDE screen

Further components of the JDE include the following:

- ❑ Compiler
- ❑ Editor
- ❑ Debugger
- ❑ Profiler

Compiler

RIM developed its compiler to optimize, compress, and obfuscate compiled files into COD format before deployment to reduce risks, reduce storage overhead on the device, and reduce transmission costs when deployed over the air (OTA). During the compilation process, images are also translated into PNG format to reduce storage requirements. If you want to port existing J2ME applications to the BlackBerry, you must first run your code through the BlackBerry compiler (see Figure 13-2). It is also possible to deploy JAR files OTA to corporate BlackBerry users, because MDS will convert J2ME JAR files automatically into COD format before they are downloaded and installed on a BlackBerry.

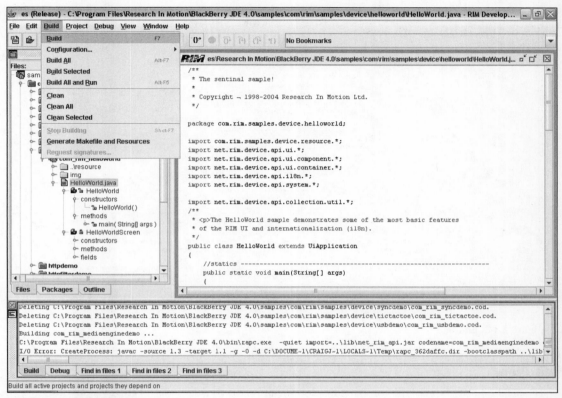

Figure 13-2: Compiler screen

Note that COD files generated by BlackBerry version 4.0 JDE work only on devices running Handheld Software version 3.8 and up. If you support earlier versions of the Handheld Software, you must also compile your application using a JDE version prior to version 4.0.

Editor

The JDE editor, shown in Figure 13-3, is designed to highlight keywords and commented sections. It can set and deselect breakpoints and symbolic bookmarks directly. It can also display methods in Outline and Packages, show class definitions, auto-complete methods, and much more.

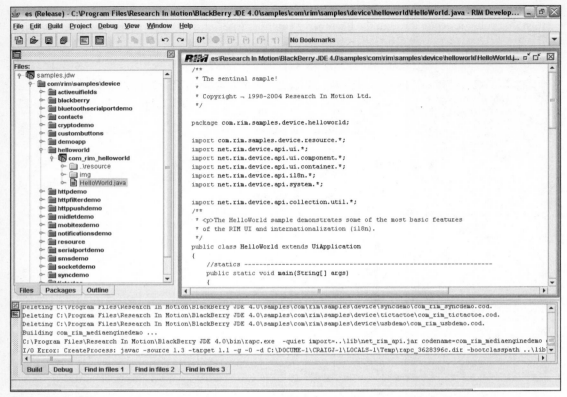

Figure 13-3: Editor

Debugger

The least-favorite facet of the software development process is usually debugging. Where designing and developing software is often a joy, finding and correcting errors is usually a tedious, problematic experience, made worse by event-driven environments in languages that rely on third-party components to work.

The debugging environment provided with the BlackBerry JDE eases much of the pain associated with tracking and eradicating bugs and other difficulties. Figure 13-4 shows the BlackBerry debugger.

The most common debugging technique is to set a few breakpoints at critical sections of your code and then gradually set breakpoints at shorter intervals to identify the source of the problem. Select Go from the Debug menu to compile and run all active applications in the simulated environment. After the program has paused at a breakpoint, use debugging tools to view various application processes and statistics to identify the problem.

A less-common technique used is to start a debugging session with the device simulator, and then click Break Now on the Debug menu in the main window. The breakpoints pane displays the section of code, and the line number at which the program is paused. The edit pane displays the source code, with an arrow indicating the line of code at which the application is paused. The output pane also displays messages from the debugger.

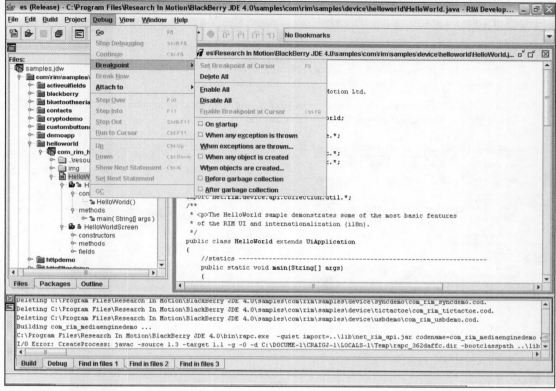

Figure 13-4: Debugger

To resume debugging after pausing at a breakpoint, click Continue on the Debug menu. To stop the debugging session in the simulator, click Quit on the File menu.

The memory statistics tool that becomes available during debugging sessions can be used to find memory leaks.

Note that debugging can also be performed on a live BlackBerry that has been connected to the computer through serial or USB ports.

Profiler

The Profiler tool shows the percentage of time spent in each code area to the current point of execution. You can use it to profile the efficiency of code sections by setting a breakpoint at the start and end of the section of code that you want to profile. You can start a debugging session with the simulator, run until the first breakpoint is reached, and then click Profile on the View menu. Click Options in the profile plane, set the profile options, and then click Go to continue running the application. When the second breakpoint is reached, click Profile on the View menu, and then click Refresh to retrieve accumulated profile data from the Java VM.

You can use profile views to display information about the section of code that you just ran. Note at the bottom of the top pane in Figure 13-5, the following tabs can be used to change views:

❑ **Summary:** This view shows general statistics about the system and garbage collector.

❑ **Methods:** This view shows a list of modules, sorted either by the information that you are profiling or by the number of times each item has been executed.

❑ **Source:** This view shows the source lines of a single method. You can navigate through the methods that call (and are called by) that method.

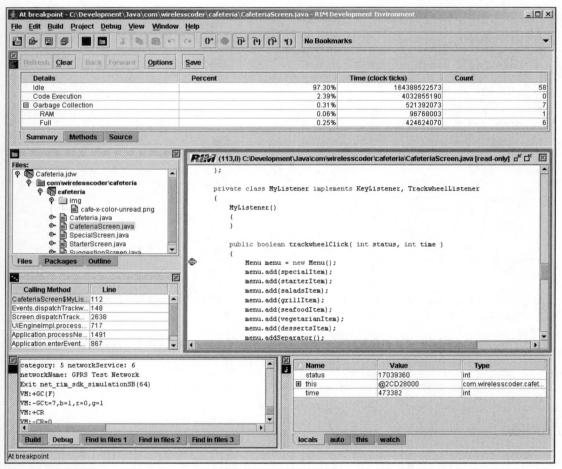

Figure 13-5: Profiler

Six options are available during a profile session, depending on the state of your profiling work:

❑ **Refresh:** Update profile from simulator.

❑ **Clear:** Clear profile statistics on simulator.

- ❑ **Back:** Profile previous method in history.
- ❑ **Forward:** Profile next method in history.
- ❑ **Options:** Profile options:
  - ❑ **Method Attribution:** Cumulative or In Method only.
  - ❑ **Sort Methods By:** Time (clock ticks) or Count.
  - ❑ **What to Profile:** Time (clock ticks), Number of objects created, Size of objects created, Number of objects committed, Size of objects committed, Number of objects moved to RAM, Size of objects moved to RAM, or User Counting.
- ❑ **Save:** Save results to file.

BlackBerry Wireless Device Simulator

The BlackBerry simulator functions similar to an actual BlackBerry wireless device, and it displays icons on the home screen for all applications that have been loaded (see Figure 13-6).

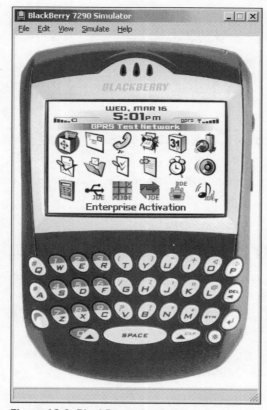

Figure 13-6: BlackBerry simulator

The simulator starts automatically when you run programs in the JDE. If it does not, access Simulator Preferences by selecting Edit ⇨ Preferences, and then select the Launch simulator check box (see Figure 13-7). If you are testing applications that require an HTTP connection, you must also start the Mobile Data Service simulator. In the JDE simulator preferences, select Launch Mobile Data Service (MDS) with simulator, or start the MDS Simulator manually.

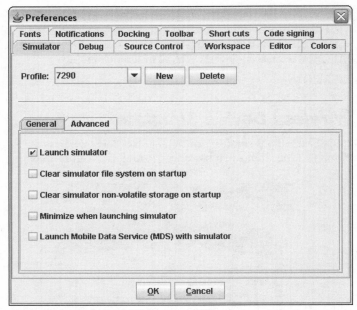

Figure 13-7: Launch simulator screen

You can start the simulator from the JDE by clicking Go on the Debug menu, or by pressing F5 on the keyboard.

Creating Your First BlackBerry Application

Creating your first application is fairly straightforward. In this section, we will examine the following:

- ❑ Preparing your workspace
- ❑ Creating and adding source files
- ❑ Developing your first application
- ❑ Fully extending the application
- ❑ Distributing the application

Preparing Your Workspace

Start the JDE, and then close down the sample workspace by using File ➪ Close Workspace.

To create a new workspace to house your project (or projects), follow these steps:

1. Select File ➪ New Workspace.
2. Enter the name of your workspace without a file extension in the Workspace name field.
3. Enter the directory path where your workspace will be stored in the Create in this directory field. For this chapter, we will use Cafeteria as our workspace name.
4. Click OK when you are ready to create the workspace.

To create a project within your workspace, follow these steps:

1. Select Project ➪ Create New Project.
2. Enter the name of your project without a file extension in the Project name field.
3. Enter the directory path where your project will be stored in the Create project in this directory field. You should follow the Java naming convention for placement of your project. For example, we called our project Cafeteria, used `www.wirelesscoder.com` as the originating domain, and stored our Cafeteria project at `c:\development\java\com\wirelesscoder\cafeteria\cafeteria.jdp`.
4. Click OK when you are ready to create your project.
5. Select Project ➪ Set Active Projects to ensure that your project is in the active list.

Creating and Adding Source Files

If you already have source files prepared, copy them into the project directory, and then add the files into your project by selecting Project ➪ Add File to Project.

Note that a default icon is used to identify your application unless you add an icon to your project and flag it as the default icon.

If you have a graphic image available to use as an icon, copy it to the project folder, or to a folder within the project folder. Afterward, add the image to your project by selecting Project ➪ Add File to Project, right-click the image after it has been added, select Properties, and then select Use as an application icon. Note that icons should be in the PNG format and should also be compressed by flattening the layers. The storage size can drop by a factor of 10 or more after the layers have been flattened.

If you want to create source files using the JDE, follow these steps:

1. Select File ➪ New.
2. Enter the filename with a `.java` extension in the Source file name field.
3. Enter the project directory path to store your file in the Create source file in this directory field.

4. Click OK when you are ready to create your file, and then type your code.

5. Save the basic template that has been created.

6. Highlight the project name in the Files window.

7. Select Project ⇨ Add File to Project.

8. Locate your file and add it.

Developing Your First Blackberry Application

For the remainder of this chapter, we will focus on developing software for the BlackBerry using the BlackBerry APIs.

A bare-bones BlackBerry application needs only a `main()` entry point to start an instance of the Event Dispatcher, a constructor to push a screen, and a screen to push. Overriding the `onClose()` method within the screen handler provides additional control over how the application terminates.

You should understand the purpose for the Event Dispatcher before starting development. When the BlackBerry is powered up, the VM is one of the first components to launch, which in turn starts an instance of the Application Manager. The Application Manager acts as the central dispatcher of operating system events for all Java applications on the device. In operation, the Application Manager receives an event and then copies it to the appropriate application queue or queues. As an example, only the current foreground application will receive user input messages, so related events must be placed in the queue of the foreground application.

Each BlackBerry application passes control to `enterEventDispatcher()` when it first starts up, which reacts to passed events that become available in its queue for the life of the application. This is the nature of an event-driven environment.

In this exercise, we'll create a basic Cafeteria application. Then we will extend it slightly to provide functionality that is similar to the Cafeteria Web application. The Cafeteria application is available at the Wrox Web site. Visit the Web site and download the file `589538 ch13 code.zip`. After you have downloaded it, unzip the contents into a directory called `c:\development`.

In our sample, we store the workspace here:

```
C:\Development\Java\Cafeteria.jdw
```

The project will be stored here:

```
C:\Development\Java\com\wirelesscoder\cafeteria\cafeteria.jdp
```

All source files were stored in the project directory. The main source file is called `cafeteria.java`.

With all Java applications, you must set your package name to match how your project was set up on the storage media.

In our example, we used a structure of `com\wirelesscoder\cafeteria\`. Therefore, the package name must be `com.wirelesscoder.cafeteria`.

```
/*
 * Filename: Cafeteria.java
 */
package com.wirelesscoder.cafeteria;
```

Next, import standard user interface and system files to make it work.

```
import net.rim.device.api.ui.*;
import net.rim.device.api.ui.component.*;
import net.rim.device.api.ui.container.*;
import net.rim.device.api.system.*;
```

It's time to write your first Cafeteria method.

All GUI applications that rely on BlackBerry APIs extend `UiApplication` and start with `main()`. The `main()` method has to pass control to the Event Dispatcher. The custom constructor takes care of making the screen appear.

```
public class Cafeteria extends UiApplication
{
    public static void main(String[] args )
    {
        Cafeteria myApp = new Cafeteria();
        myApp.enterEventDispatcher();
    }

    public Cafeteria()
    {
        pushScreen(new CafeteriaScreen());
    }
}
```

Save this file and then add it to your project by selecting Project ⇨ Add File to Project. Create a new file called `cafeteriascreen.java` by selecting File ⇨ New. Although you can include the primary screen handler within the main source file, it is better to separate individual components into their own source files to make it easier to maintain and modify them.

We start by creating a screen handler that extends `MainScreen`. It starts with a constructor that simply creates something to view and then displays it. We also override the `onClose()` method to query the user to see if the user really wants to exit the program.

```
/*
 * Filename: CafeteriaScreen.java
 */
package com.wirelesscoder.cafeteria;

import net.rim.device.api.ui.*;
import net.rim.device.api.ui.component.*;
import net.rim.device.api.ui.container.*;
import net.rim.device.api.system.*;

final class CafeteriaScreen extends MainScreen
{
```

```
    public CafeteriaScreen()
    {
        super();
        LabelField title = new LabelField("Cafeteria",
          LabelField.ELLIPSIS | LabelField.USE_ALL_WIDTH);
        setTitle(title);
        add(new RichTextField("Company X's Cafeteria Portal"));
    }

    public boolean onClose()
    {
        if ( Dialog.ask(Dialog.D_YES_NO, "Exit Application?") == Dialog.YES )
        {
            System.exit(0);
            return true;
        }
        return false;
    }
}
```

Two areas within this source module are of interest.

❑   `add(new RichTextField("Company X's Cafeteria Portal"));`: Creates a RichText field to display data on the screen.

❑   `if ( Dialog.ask(Dialog.D_YES_NO, "Exit Application?") == Dialog.YES )`: Used in the `onClose()` method to query the user before allowing the application to terminate.

Save this source file and then add it to the project. Ensure that debugging files are created during compile by selecting Build ⇨ Configuration ⇨ Debug. Build your application by selecting Build ⇨ Selected, and try it out by selecting Debug. It will provide the framework you will need to start creating your own applications.

Fully Extending the Application

In this exercise, we will replace `cafeteria.java` with a version that adds and handles new menu items and calls three external screens.

The opening verbiage remains the same:

```
/*
 * Filename: CafeteriaScreen.java
 */
package com.wirelesscoder.cafeteria;

import net.rim.device.api.ui.*;
import net.rim.device.api.ui.component.*;
import net.rim.device.api.ui.container.*;
import net.rim.device.api.system.*;
```

The constructor is changed to set up listeners for Key and Trackwheel events:

```
final class CafeteriaScreen extends MainScreen
```

```
{
    public CafeteriaScreen()
    {
        super();
        LabelField title = new LabelField("Cafeteria",
            LabelField.ELLIPSIS | LabelField.USE_ALL_WIDTH);
        setTitle(title);
        add(new RichTextField("Company X's Cafeteria Portal"));
        MyListener listener = new MyListener();
        this.addKeyListener(listener);
        this.addTrackwheelListener(listener);
    }
```

onClose() is changed to directly exit the application:

```
    public boolean onClose()
    {
        System.exit(0);
        return true;
    }
```

We insert a handler for the menu items Specials and Starters that push external screens to the display.

```
    private MenuItem specialItem = new MenuItem("Specials", 100, 10 )
    {
        public void run()
        {
            UiApplication.getUiApplication().pushScreen(new SpecialScreen());
        }
    };

    private MenuItem starterItem = new MenuItem("Starters", 110, 10 )
    {
        public void run()
        {
            UiApplication.getUiApplication().pushScreen(new StarterScreen());
        }
    };
```

The "Salads", "Grills", "Seafood", "Vegetarian" and "Desserts" menu items are
handled very simply by displaying a Dialog box.

```
    private MenuItem saladsItem = new MenuItem("Salads", 120, 10 )
    {
        public void run()
        {
            Dialog.inform("Today's salads");
        }
    };

    private MenuItem grillItem = new MenuItem("Grills", 130, 10 )
    {
        public void run()
        {
            Dialog.inform("Today's grill");
```

```
        }
    };

    private MenuItem seafoodItem = new MenuItem("Seafood", 140, 10 )
    {
        public void run()
        {
            Dialog.inform("Today's seafood");
        }
    };

    private MenuItem vegetarianItem = new MenuItem("Vegetarian", 150, 10 )
    {
        public void run()
        {
            Dialog.inform("Today's vegetarian");
        }
    };

    private MenuItem dessertsItem = new MenuItem("Desserts", 160, 10 )
    {
        public void run()
        {
            Dialog.inform("Today's desserts");
        }
    };
```

Note that you must reference the main instance of the application to push a screen through:

```
UiApplication.getUiApplication().pushScreen(new SCREEN_HANDLER());
```

Also note that each menu handler takes three parameters: name, ordinal (sequence within the menu), and priority. For example, the following snippet creates a menu item called Desserts that has an ordinal of 160 and a priority of 10:

```
private MenuItem dessertsItem = new MenuItem("Desserts", 160, 10 )
```

The Suggestion Form is also displayed using a separate screen handler:

```
    private MenuItem suggestionItem = new MenuItem("Suggestion Form", 170, 10 )
    {
        public void run()
        {
            UiApplication.getUiApplication().pushScreen(new SuggestionScreen());
        }
    };
```

A new Close menu handler is introduced as the last menu item that will prompt the user on exit. It calls onClose() if the user chooses to exit the application:

```
    private MenuItem closeItem = new MenuItem("Close", 200000, 10 )
    {
        public void run()
```

```
        {
            if ( Dialog.ask(Dialog.D_YES_NO, "Exit Application?") == Dialog.YES )
            {
                onClose();
            }
        }
    };
```

A listener class (`MyListener`) is created to handle `Key` and `Trackwheel` events. The `trackWheelClick` event is used to add menu items, and the `keyChar` event is used to exit the application.

```
        private class MyListener implements KeyListener, TrackwheelListener
    {
        MyListener()
        {
        }


        public boolean trackwheelClick( int status, int time )
        {
            Menu menu = new Menu();
            menu.add(specialItem);
            menu.add(starterItem);
            menu.add(saladsItem);
            menu.add(grillItem);
            menu.add(seafoodItem);
            menu.add(vegetarianItem);
            menu.add(dessertsItem);
            menu.addSeparator();
            menu.add(suggestionItem);
            menu.addSeparator();
            menu.add(closeItem);
            menu.show();
            return true;
        }

        /**
        * Invoked when the trackwheel is released
        */
        public boolean trackwheelUnclick( int status, int time )
        {
            return false;
        }

        /**
        * Invoked when the trackwheel is rolled.
        */
        public boolean trackwheelRoll(int amount, int status, int time)
        {
            return false;
        }

        /**
        * Invoked when a key is pressed.
        */
```

```
            public boolean keyChar(char key, int status, int time)
            {
                if (key == Characters.ESCAPE)
                {
                    onClose();
                    return true;
                }
                return false;
            }

            /** Implementation of KeyListener.keyDown */
            public boolean keyDown(int keycode, int time)
            {
                return false;
            }

            /** Implementation of KeyListener.keyUp */
            public boolean keyUp(int keycode, int time)
            {
                return false;
            }

            /** Implementation of KeyListener.keyRepeat */
            public boolean keyRepeat(int keycode, int time) {
                return false;
            }

            /** Implementation of KeyListener.keyStatus */
            public boolean keyStatus(int keycode, int time) {
                return false;
            }
        }
    }
```

Save the file and then create a new screen handler called SpecialScreen.java.

This handler is extremely simple. It displays a title, and then displays each Cafeteria special in a separate RichTextField. Note that formatting is permitted within RichTextFields, so line feeds are included to break up the text.

```
/*
 * Filename: SpecialScreen.java
 */
package com.wirelesscoder.cafeteria;

import net.rim.device.api.ui.*;
import net.rim.device.api.ui.component.*;
import net.rim.device.api.ui.container.*;
import net.rim.device.api.system.*;

final class SpecialScreen extends MainScreen
{
    public SpecialScreen()
    {
```

```
        super();
        LabelField title = new LabelField("Cafeteria Specials",
            LabelField.ELLIPSIS | LabelField.USE_ALL_WIDTH);
        setTitle(title);
        add(new RichTextField("Surf and Turf\n- Sirloin served with pan fried or
deep
            fried calamari with lemon butter, peri-peri or garlic sauce\n"));
        add(new RichTextField("Crab and Mushroom Steak\n- Sirloin covered in a
creamy
            crab and mushroom sauce\n"));
        add(new RichTextField("Jalapéno Steak\n- Sirloin stuffed with jalapéno's
covered
            in a creamy peri-peri sauce\n"));
        add(new RichTextField("Carpetbagger Fillet\n- Stuffed with smoked
mussels\n"));
        add(new RichTextField("Mushroom and Fillet Stack\n- Medallions of fillet
layered
            with black mushrooms, melted mozzarella cheese and topped with mushroom
            sauce"));
    }
}
```

Save the file and add it to your project. Then create a new screen handler called StarterScreen.java.

This handler works the same way as SpecialScreen.java:

```
/*
 * Filename: StarterScreen.java
 */
package com.wirelesscoder.cafeteria;

import net.rim.device.api.ui.*;
import net.rim.device.api.ui.component.*;
import net.rim.device.api.ui.container.*;
import net.rim.device.api.system.*;

final class StarterScreen extends MainScreen
{
    public StarterScreen()
    {
        super();
        LabelField title = new LabelField("Cafeteria Starters",
            LabelField.ELLIPSIS | LabelField.USE_ALL_WIDTH);
        setTitle(title);
        add(new RichTextField("Calamari\n- Pan fried in either garlic, peri-peri or
lemon
            butter or deep fried and served with tartare sauce and savoury
rice\n"));
        add(new RichTextField("Mussels\n- Done in a creamy white wine garlic
sauce\n"));
        add(new RichTextField("Potato Skins\n- Deep fried with melted mozz-cheddar
and
            bacon bits\n"));
```

```
        add(new RichTextField("Calamari Steaks\n- Cut into strips and deep fried or
pan
            fried in either garlic, lemon butter or peri-peri sauce"));
    }
}
```

Save the file and add it to your project. Then create a new screen handler called
SuggestionScreen.java.

This handler is more complex than the previous two. An edit field is provided to enable the user to type a name, and two option buttons are displayed for the user to select Yes or No.

```
/*
 * Filename: SuggestionScreen.java
 */
package com.wirelesscoder.cafeteria;

import net.rim.device.api.ui.*;
import net.rim.device.api.ui.component.*;
import net.rim.device.api.ui.container.*;
import net.rim.device.api.system.*;

final class SuggestionScreen extends MainScreen
{
    private EditField editName;
    private RadioButtonField rbYes;
    private RadioButtonField rbNo;

    public SuggestionScreen()
    {
        super();
        LabelField title = new LabelField("Cafeteria Portal Suggestion Form",
            LabelField.ELLIPSIS | LabelField.USE_ALL_WIDTH);
        setTitle(title);
        editName = new EditField("Name: ", "", 20,
            EditField.FILTER_DEFAULT | EditField.NO_NEWLINE);
        add(editName);
        add(new RichTextField("\nDo you think that this portal will be useful to
you?"));

        RadioButtonGroup rbGroup = new RadioButtonGroup();
        rbYes = new RadioButtonField("Yes");
        rbNo  = new RadioButtonField("No");
        rbGroup.add(rbYes);
        rbGroup.add(rbNo);
        add(rbYes);
        add(rbNo);
    }

    protected boolean onSavePrompt()
    {
        if ( Dialog.ask(Dialog.D_YES_NO, "Save and exit?") == Dialog.YES )
        {
            //
```

```
                    // insert whatever code you want to save the form in persistent store
                    // email the form, or submit it to a web site via an HTTP request
                    //
                    // editName.getText();
                    // rbYes.getState();
                    // rbNo.getState();
                    return true;
            }

            //
            // keep working on the form
            //
            return false;
        }
    }
```

We overrode the onSavePrompt() method to prompt the user when attempting to leave after modifying the form. Nothing of consequence is done with the data if the user opts to save the data. You could alter the application to send the data as an e-mail message, submit it to a Web site through an HTTP post, or store the data in persistent store. We chose not to enable additional functionality to keep the application simple and non-signed because e-mail, HTTP requests, and persistent storage are handled through secure APIs. If you developed signed applications, you can use the simulator to test without signing, but will need to purchase signature keys from RIM before deployment.

Save the file and add it to your project. Build the project and test it using the device simulator. Note that the main screen is very basic and relies on menu items for navigation. Use the device's track wheel to activate the menu.

The basic framework of this application is good enough for you to develop a wide range of BlackBerry applications. Be sure that you rely heavily on the API Reference that has been provided with the JDE. While terse in nature like most Javadoc files, it does provide solid details about all APIs directly from the source code.

Distributing Your New Application

There are currently three ways in which you can distribute your application. Before we distribute the application using one of the methods listed in this section, we must first prepare and build the application. To do this, click the Build menu, scroll down to Configuration, and choose Release (see Figure 13-8).

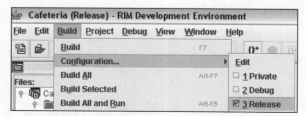

Figure 13-8: Choosing the Release option

The next step is to build your project. Click the Build menu and choose Build Selected (see Figure 13-9).

The final preparation step is to create an ALX file (see Figure 13-10). An ALX file is used by the BlackBerry Desktop Manager or Application Loader to load an application onto your handheld. The ALX file tells the Desktop Manager or Application Loader information about the application and where to find the files to load onto the handheld.

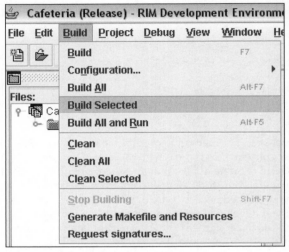

Figure 13-9: Build Selected project

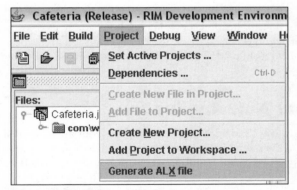

Figure 13-10: Build an ALX file

The ALX file will be placed in the same directory as your project files. Our ALX file for the Cafeteria Portal looks like this:

```
<loader version="1.0">
    <application id="com_wirelesscoder_cafeteria">
        <name >
            Cafeteria
        </name>
        <description >
```

```
            Sample application
        </description>
        <version >
            1.0
        </version>
        <vendor >
            MyCompany
        </vendor>
        <copyright >
            Copyright (c) 2005 MyCompany
        </copyright>
        <fileset Java="1.18">
            <directory >
            </directory>
            <files >
                com_wirelesscoder_cafeteria.cod

            </files>
        </fileset>
    </application>
</loader>
```

Now that you have prepared your project for distribution you can use one of the following methods:

❑   The first is the traditional method of using the BlackBerry Desktop Manager. The application files are copied to the desktop computer (either through a simple file copy or through a setup program) and added through the Application Loader. For our Cafeteria application, we must copy the ALX file and the COD file to the PC. When we run the Application Loader, we will choose to add applications to the handheld, and then browse to the ALX file. Using the ALX file, the Application Loader will install the COD file onto the handheld.

❑   The second method is to place your compiled files on a Web server and provide your users with the URL. When the users connect to the URL, the BlackBerry will display the attributes of the application and provide a download button. After it has been pressed, the application is downloaded to the handheld. Before you make the application available to your users, you must prepare your Web server by adding the following MIME types:

```
- MIME Type: application/vnd.rim.cod Extension: cod
- MIME Type: text/vnd.sun.j2me.app-descriptor Extension: jad
```

When adding MIME types to a Web server, you normally have to restart it for the changes to take effect. After your Web server has restarted, create a simple download page that links to the application. A page similar to this will surface:

```
<html>
  <body>
    <a href="com_wirelesscoder_cafeteria.jad">
      Download Cafeteria Application
    </a>
  </body>
</html>
```

Now, copy over the JAD and COD file from your project directory to complete the Web server setup. Browse to the page `http://localhost/cafe-app.htm` on your BlackBerry Simulator.

Make sure that you are running the MDS Simulator before you open this Web page. You will see a simple download Web page (see Figure 13-11). Click the link, and the BlackBerry displays the information about the application and enables you to download it (see Figure 13-12). After you click the Download button, the application is downloaded to the handheld. Once it has been downloaded, it will be installed (see Figure 13-13).

❑ The final method is one that was introduced with the BlackBerry 4.0 platform. As long as your BES is 4.0 and the Handheld Code on your handheld units is 4.0, you will be able to push out your application using software configurations. Refer to Chapter 5 for more information on how to distribute applications wirelessly through BES 4.0.

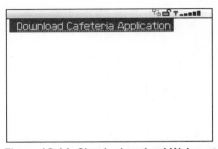

Figure 13-11: Simple download Web page

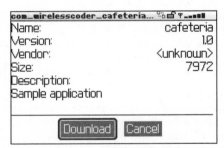

Figure 13-12: Application download screen

Figure 13-13: Application installed successfully

Summary

Programming for the BlackBerry (or any mobile device in J2ME) can really provide the end user with an enhanced mobile computing experience. As a programmer, you can add enjoyment through games, and added productivity through your applications. As mobile computing becomes more and more popular, and as the growth of the BlackBerry user population explodes, your code can be very useful.

If you take advantage of the BlackBerry-specific APIs you can really add benefit to the BlackBerry experience. This chapter discussed the writing of a BlackBerry application by creating a Cafeteria application similar to the Cafeteria portal that we created in Chapter 10.

In Chapter 14, we will discuss the Plazmic Media Engine, which enables you to enhance your BlackBerry users' Web browsing experiences.

14

The Plazmic Media Engine

The Plazmic Media Engine (PME) is software that is pre-loaded onto all BlackBerry devices starting with release 3.7 of the Handheld Code. The PME can also be loaded onto other mobile devices, and content you create using Plazmic can run across multiple mobile platforms.

The Plazmic Content Developer's Kit (CDK) is a set of tools and utilities that enables you to create rich Web content for mobile handheld devices without taking up large amounts of memory. The Plazmic CDK enables you to create content that has vibrant graphics, animation, and audio. To reduce memory and provide overall better quality graphics on the handheld device, Plazmic uses vector graphics as opposed to bitmaps.

The discussion in this chapter examines the following:

❑ Introducing the PME

❑ Creating Plazmic content

❑ Working with Composer

Introducing the PME

In previous chapters, we demonstrated how you can create Web content that is easily viewable on the BlackBerry through the BlackBerry Browser. There are many things that you can do to make the content dynamic and interesting, including the use of small bitmap graphic files. You cannot however, animate content or otherwise spice up the look and feel of the Web content. The Plazmic CDK, along with the PME, enables you to do this. Since revision 3.7 of the Handheld Code, the PME has been included on the BlackBerry, which has provided your BlackBerry with the ability to display Plazmic content.

Plazmic content approaches graphics in a completely different way, by using vector graphics instead of bitmaps. Bitmaps are basically image files made up of pixels. Each pixel has a position and a color associated with it. If you enlarge a bitmap image, you will find that it becomes jagged, or pixelated. This is because each pixel is simply being enlarged. There is no information in the bitmap image that tells the graphic program how to enlarge the image.

Vector graphics, however, are created using vectors. This technology has been around for a very long time. Some of you may remember learning how to program on an Apple II using the Apple Logo program. The idea was that you had a turtle to which you issued directional instructions and it moved around on the screen to draw a picture. You could put these instructions into a file and play it back to re-create your picture. You could tell the turtle to go up, down, left, right, and so on. Another example of vector graphics can be found in the original Atari Asteroids arcade game. The graphics in Asteroids are actually vector graphics.

Vector graphics are made up of instructions on how to draw the image. This lends itself well to resizing because there are no absolute pixel placements, but rather directional instructions. So, if you want to double the size of the image, the program simply doubles the length of the vectors. Vector graphic files are much smaller than conventional bitmap files, because there is no need to map each pixel. There are simply directional instructions.

Plazmic uses vectors to draw the images, but it also uses them to control where they move on the screen when animation is used. The vectors plot where the images will move, how quickly, how many times, and so on. Finally, Plazmic enables you to play Musical Instrument Digital Interface (MIDI) files together with the rich animated content to enhance the user experience.

RIM has a Web site that is created using the Plazmic CDK. The site is a great example of what you can achieve with the product. From your BlackBerry (or BlackBerry Simulator), browse to `http://mobile.blackberry.com` to see the site, shown in Figure 14-1.

Figure 14-1: The RIM Plazmic site

Creating Plazmic Content

To create Plazmic content you will first need to download the Plazmic CDK from www.plazmic.com/en/
download/index.shtml. The creation of Plazmic content can be broken up into three stages:

❑ **The content building stage:** This building process consists of creating images, animating them,
and creating the general look and feel of the page. The CDK includes a Graphical User Interface
(GUI) tool called Composer that enables you build your project, as shown in Figure 14-2. The
Composer has a built-in BlackBerry Simulator, shown in Figure 14-3, that enables you to see
how your project will look on a real BlackBerry. The Composer enables you to import Scalable
Vector Graphics (SVG) files, which are standard vector graphics files that you may have created
using other tools in the past. After your project is built, you can export it from Composer. When
your project is exported, it is saved along with the images that you used within the project.

❑ **The deployment stage:** You can use either the Composer tool or the included SVG Transcoding
Utility to convert your project to a compressed binary file that is ready for viewing on the handheld device. The compressed binary file has the .pme extension. To further reduce network traffic,
you can bundle the images used in the project into one file with a .pmb extension. When you have
a .pme or .pmb file, you can post it to a Web server ready for viewing.

❑ **The using stage:** The final stage is to run the content on the handheld device. Since the
BlackBerry Browser has built-in support for Plazmic, the user simply points the browser to it.

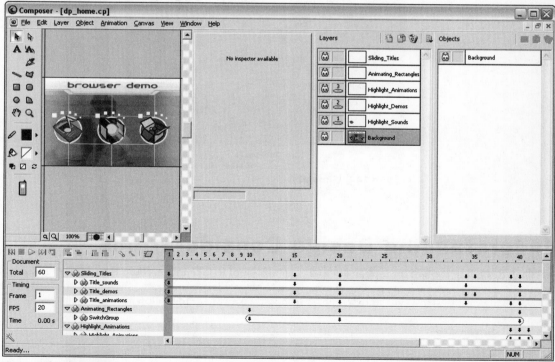

Figure 14-2: Composer

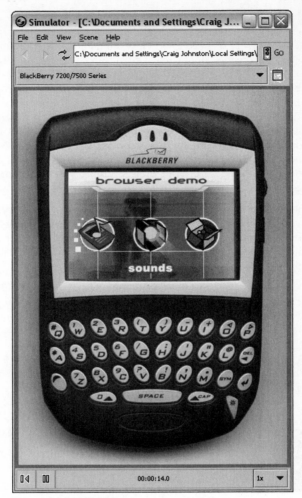

Figure 14-3: Simulator

Understanding Design Considerations

When designing content for mobile devices, it is important to take into account their limitations. As mobile devices become more powerful and wireless networks become faster, these limitations are decreasing rapidly. However, at this time, you should be aware of the following:

- ❑ Screen size
- ❑ Color depth
- ❑ Fonts
- ❑ Sound

Screen Size

The different models of BlackBerry handheld units have differing screen sizes. For example:

- **BlackBerry 6200 series:** 160×160 pixels
- **BlackBerry 7200 series:** 240×160 pixels
- **BlackBerry 7700 series:** 240×240 pixels
- **BlackBerry 7100 series:** 240×260 pixels

Color Depth

The monochrome BlackBerry devices are true monochrome and their screens are not grayscale. This means that any content you create for the monochrome handheld units must not use colors of any kind, because they will look very unattractive in monochrome. The color BlackBerry devices have 16-bit color screens, which means that they can display 65,536 colors.

Fonts

The BlackBerry devices support the BlackBerry Millbank font while the Plazmic CDK uses the Tahoma font by default, which looks similar in appearance to the BlackBerry Millbank font. Any fonts that the handheld unit does not support are converted to bitmaps when your content is exported from the Composer tool.

Sound

The BlackBerry handheld units support MIDI files. Single-track, single-instrument MIDI files will sound best on the BlackBerry. If the MIDI files have multiple instruments, instrument 0 will be selected. Instrument 0 is normally the piano. All five types of MIDI files are supported, including PPQ, SMPTE_24, SMPTE_25, SMPTE_30DROP, and SMPTE_30. When creating MIDI content, try to follow these rules:

- Create single-track, single-instrument files
- Avoid simultaneous or overlapping notes
- Use an octave range between four and seven (MIDI numbering)
- Avoid MIDI commands within the file

If the MIDI files do not play correctly on the BlackBerry, try setting the sequencing to one octave higher or decreasing the playback resolution or pulses per quarter note.

Working with Composer

Now that we have a brief understanding of the Plazmic content-creation process, as well as the rules and recommendations surrounding creating mobile content, let us start at the beginning and learn how to use the Composer tool.

Creating New Content

When you start Composer and select File ➪ New, you are presented with a dialog box that shows different templates (see Figure 14-4). These templates represent the different models of BlackBerry. You can modify the templates and create new ones when new BlackBerry devices are released. For example, the BlackBerry 7100 series has a screen resolution of 240 × 260 pixels, so you can create a new template for that device if you plan to create content for it. In Figure 14-5 you can see that modifying a template involves giving it a name, specifying the width and height of the canvas (or screen), and choosing the background color.

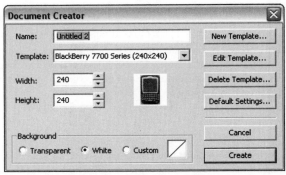

Figure 14-4: Creating a New Composer document

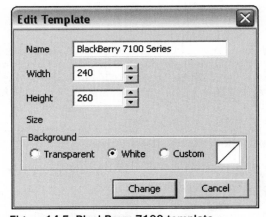

Figure 14-5: BlackBerry 7100 template

You can also open existing content with Composer. The Composer supports the following file types:

❑ SVG

❑ Bitmap (.gif, .jpg, .png, .tif, .bmp, .tga, .pme, .pmb)

❑   Adobe Illustrator (version 8 or earlier)

❑   Beatware e-Picture Pro (.ep)

While working on your project, you can import SVG files and the following bitmap files into it:

❑   `.3ds`

❑   `.bmp`

❑   `.dxf`

❑   `.gif`

❑   `.jpg`

❑   `.lwo`

❑   `.png`

❑   `.psd`

❑   `.tga`

❑   `.tif`

❑   `.wmf`

After you have opened an existing Composer file or created a new one, you will be presented with a number of windows within Composer. Each window has a specific purpose and enables you to work with different aspects of your document.

Document Window

The document window displays your work as you build your project. The *Canvas* is the area of your project where all of the action will take place. Think of the Canvas area as the stage for your animation and other content to play out. The Canvas size is determined by your template. In the case of a BlackBerry, those sizes are predetermined by the BlackBerry model.

The Overscan area is the area of your project that is off-stage. You will use this area to place your objects before they move into the Canvas area (or stage) to perform.

If you open the sample file called `sub_demos_skiing` (which you will find in `c:\program files\ plazmic cdk 3.7\samples`), you will see how the Canvas and Overscan areas are used (see Figure 14-6). Before the animation starts, you can see that the Canvas area (or stage) is black with two icons in the lower-left corner. In the Overscan area (or offstage), you can see objects waiting to come onstage to perform. We see two skiers and a group of white objects that will act as snow.

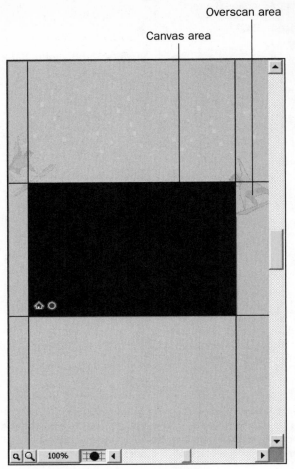

Figure 14-6: Canvas/overscan example

When you have created a new document, you can modify the Canvas by selecting Canvas ⇨ Properties. To select the Canvas properties, right-click anywhere in the Document window and choose Properties (see Figure 14-7). On this screen, you can change the color and size of the Canvas.

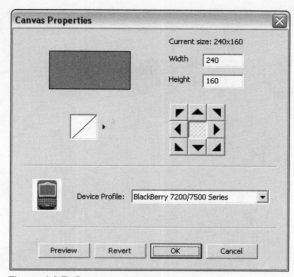

Figure 14-7: Canvas properties

Multiple Views

Composer enables you to display multiple views of the same project. For example, in one view, you could show the project in full 16-bit color, while in another you could show it with a reduced color depth. In another view, you may want to show the project zoomed, to pick out certain aspects of your work.

To show multiple views of the same project, first ensure that the document window is not maximized. If it is, click the restore icon in the top right of the screen. Then click the View menu and choose New Viewer. A new view of the same project will appear, as shown in Figure 14-8. Each view is treated independently. You can change the way each view displays the document by selecting the view window and making changes to the view (such as zooming in or out, or changing the color depth).

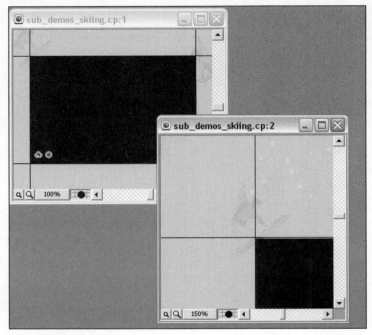

Figure 14-8: Multiple views

Grids and Guides

Like many graphics programs, Composer has grids and guides to help you place objects within your project. When you show the grids, a grid pattern starts at the top left of the document window. The grid lines are spaced equally across and down.

To show the grid lines, click the View menu and choose Show Grid. To modify the grid spacing, click the View menu and choose Set Grid Options to display the screen shown in Figure 14-9. This enables you to change the grid spacing and color of the grid lines.

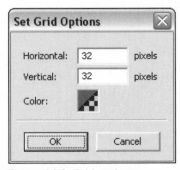

Figure 14-9: Grid options

You can use these grid lines as markers to make lining up objects in your project easier. Alternatively, if you want Composer to help you place these objects, you can enable the Snap to Grid feature. This causes Composer to snap your objects to the grid lines to accurately place them.

If you want to be more granular about placing your objects, you can use Guides. To enable Guides, click the View menu and choose Guides. Next, click the View menu and choose Show Rulers. After these menu options have been selected, you can click inside the rulers and drag your mouse onto the canvas. A guide line will follow your mouse. This enables you to place guide lines anywhere you like. Like the grid lines, you can enable the Snap To Guides feature to let Composer snap objects to the guide lines. To do this, click View and choose Snap To Guides, as shown in Figure 14-10.

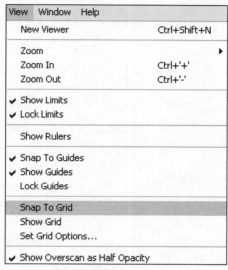

Figure 14-10: Using guides

Objects and Layers

Objects are items that you can create, modify, and animate within your project. You can group objects in *layers* within your project. Using layers enables you to animate objects on their own plane of existence, completely separate from other objects that are on other layers. To create objects you will use the Toolbox (see Figure 14-11).

The following table describes the Toolbox tools. The order of presentation in the table matches the icons in Figure 14-11, reading from left to right and top to bottom.

Figure 14-11: Toolbox

Tool	Description
Arrow	Select and move objects. In addition, change the width and height of objects.
Transformation	Reshape an object by grabbing and moving a corner. You can use this tool on lines, polylines, ellipses, rounded rectangles, arcs, beziers (curved lines), and rectangles.
Bezier	Draw curved lines.
Text	Create a text object.
Text on Curve	Create text on a curve.
Line	Draw a straight line.
Polyline	Draw a line made up of multiple segments.
Rectangle	Draw a rectangle.
Rounded Rectangle	Draw a rectangle with rounded corners.
Ellipse	Draw a circle or elliptical object.
Arc	Draw an arc.
Hand	Use the hand to move the document window around.
Zoom	Magnify the document window.
Stroke	Select the stroke color or color of the lines that make up the object.
Fill	Select the fill color or the color that fills the inside of the object.

Tool	Description
Swap Fill and Stroke color	Use this to swap the fill and stroke color.
Default Colors	Select the default colors.
No color	Use this to remove the fill color.
Simulator	Launch the Simulator and show the current project.

Use one of the tools to create an object. To create a new object, click the object type in the Toolbox. With the tool selected, click and drag your mouse on the canvas. This places the object and creates its initial size. While the object is selected, you can use the Inspector window (which is a window within the Composer tool) to modify the object's parameters. The Inspector window enables you to change all aspects of objects within your document. Sometimes, the Inspector tool is hidden behind other windows, so you may have to move the different windows around to find it. If you are creating a Bezier, polyline, text, or text on a curve, you must press the Enter key to end the object creation. You can then use the Inspector window to modify the object's parameters.

The one object that differs slightly in the way in which it is created is the Text on Curve object. To create text on a curved object, you must first create an object that has the shape you want for your text. For example, you can create an elliptical object. Next, click the Text on a Curve tool. Then, click the object around which you want to wrap the text. The text entry icon appears, enabling you to type your text. When you are finished, press Enter. When you click outside the object, the original shape disappears, displaying only the curved text.

As you create your project you may have multiple objects on multiple layers. You can use the Layers window, shown in Figure 14-12, to display or hide your different layers. This makes it easier to select objects, especially if they are overlapping. To do this, click the Eyes icon (in the lower-left of Figure 14-12) on each layer to cause the layer to display or be hidden. The Layers window is normally to the right of the Document window.

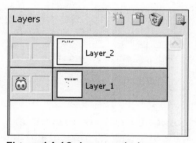

Figure 14-12: Layers window

Anti-Aliasing

In the days of analog Cathode Ray Tubes (CRT), curved objects and slanted lines did not look jagged because of the imprecise nature of the analog display. In a sense, the display was doing accidental *anti-aliasing* (smoothing out of the lines). When images are viewed on a Liquid Crystal Display (LCD) like the screen on a BlackBerry, they are displayed precisely. This causes curves and slanted lines to have a jagged appearance. To cause the appearance of smoothed edges, anti-aliasing tricks the eye into seeing a smooth edge. In Composer, you can enable anti-aliasing on any object so that when it displays on the LCD screen, it looks like it has smooth edges.

Anti-aliasing is not supported on vector objects when they are displayed on the BlackBerry. So, if you must enable anti-aliasing on a vector object, you must first convert it to a bitmap object, or *rasterize* it. To do this, select the object on the Canvas, then on the menu bar select Object ⇨ Rasterize.

To enable anti-aliasing for a particular object, select the object on the canvas and select the Anti-aliasing check box in the Inspector window.

Filters

Filters can be used to artistically display bitmap objects. To apply one or more filters to a bitmap object, select the object on the Canvas, and click the Filters tab in the Inspector window (see Figure 14-13). You can even animate filters. We have not yet touched on animation, but when we do, we will cover how to animate filters. The Filters tab is the third tab from the left in the Inspector window.

Figure 14-13: Filters tab

Effects

Effects can be applied to bitmap objects to modify their appearances. While an object is selected on the Canvas, click on the Effects tab in the Inspector window to add and modify the effects (see Figure 14-14). The Effects tab is the fourth tab from the left in the Inspector window.

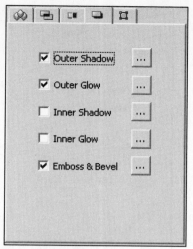

Figure 14-14: Effects tab

Animation

If you think of animated films, you typically think of stop-motion animation or cartoon animation. Both of these techniques are considered frame-by-frame animation where each frame of animation is recorded. *Stop-motion animation* is where an object being animated is posed for each frame of a film. So, for a 1-second shot (which consists of 24 to 30 frames), poses must be struck by the model, object, or whatever is being animated. For *cartoon* shows, the animation is the same, but this time a character is drawn in a specific position for each frame of film.

Composer uses *keyframe animation*, which differs from frame-by-frame animation in the way that objects are animated. Keyframe animation reduces the size of the animation files, which is more efficient when transmitting them across low-bandwidth networks.

When you construct animation using keyframe animation, you animate an object by specifying how the object looks, and what position it is on the screen at each keyframe. Composer interpolates the object between the keyframes to create a fluid animation. This means that most of the animation work is being done by the PME.

Creating Your First Animation

To see how keyframe animation works, let us create a simple animation. Start the Composer, change the template to the BlackBerry 7700 series by selecting it in the Template drop-down field. Call your document Anim1. Ensure that the background is white and click Create.

We will create a simple shape and have it move from one location on the screen to another. To do this, create a rounded rectangle anywhere on the canvas. Click the Rounded Rectangle icon in the Toolbox. Click and drag your mouse on the canvas to place and size the object. While the object is selected, click the Attributes tab of the Inspector window and change both Translate values to 170. Change the Dimension to 40 and change the color to orange, as shown in Figure 14-15.

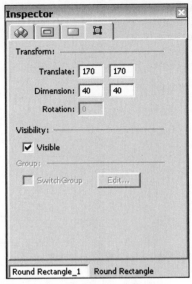

Figure 14-15: Object attributes

On the bottom of the screen in the Animation window, you will see that one frame has been added to the animation. It shows the object that you have just created called Round Rectangle_1. As you add more objects to your animation, you may want to rename the objects to something meaningful like sun, moon, and so on, to represent what they are in the animation.

On the bottom left of the Animation window, click the Animation Wizard icon. On the next screen, select Insert Frames. When you click Next, change the Parameters to Insert **49** frames **after** frame **1**, as shown in Figure 14-16.

Figure 14-16: Frame wizard

When you click Finish, the wizard will add the frames to your document. Now that your frames have been added, click the arrow next to the object Round Rectangle_1 (see Figure 14-17). This will show the four animation types that can be done between keyframes:

❏ **Transition:** How the object moves between keyframes

❏ **Visibility:** How visible the object is

❏ **Color - Stroke:** The color of the outline of the object

❏ **Color - Fill:** The color of the inside of the object

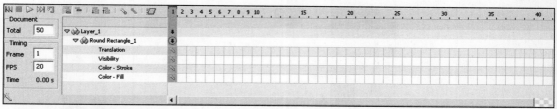

Figure 14-17: Animation window

In Figure 14-17, you will see the layout of the Animation window. On the top of the window starting on the left-hand side, you will see buttons to control the playback of animations. To the right of the play controls, you will see buttons to control which items are displayed in the Animation window and whether to display the content that has subcategories expanded out or not. To the right of that are the buttons to create and delete keyframes. Next to that is a button to create a new Timeline Action, which we will discuss later in the chapter. The next section shows the frame numbers that make up your animation along the top, while the objects within your animation are shown in a list to the left frame 1.

You will notice the key icons next to each animation type. Key icons represent events that occur at a particular frame in the animation. This is the heart of keyframe animation.

What we want to do is move our rounded rectangle from its current position to one that is higher on the screen, and change its color. The idea is to simulate a rising sun. To do this, scroll over and click frame 50 (see Figure 14-18). Then click the Add Key icon and click Translation line (not the word Translation). When you added the key to frame 50, the Animation window automatically displays the attributes that can be changed for the rounded rectangle object. We will start by changing the Translation or position of the object.

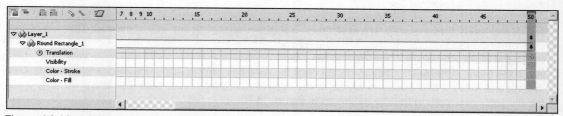

Figure 14-18: Added Translation keyframe

While frame 50 is selected, drag the Rounded Rectangle on the canvas to a new spot. What you are really doing is positioning the object in the location that it must be in frame 50, as shown in Figure 14-19.

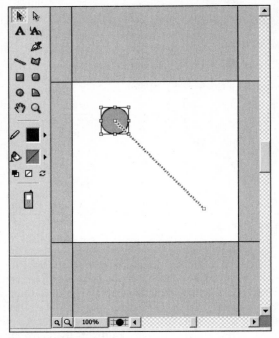

Figure 14-19: Moved sun

If you click the Play button in the Animation window, you will see your sun rising in the sky.

To make it more authentic, we can change the color of the sun to a bright yellow in frame 50 so that it slowly becomes brighter as it animates. To do this, select frame 50 and modify the attributes of the object. Change the color of your rounded rectangle to bright yellow. Now when you click the Play button, your sun will rise and get brighter. If you want to make it even more authentic, change the properties of the Canvas to a blue color.

To see how this will look on the BlackBerry, click the Simulator button in the document window. When you are finished, close the Simulator.

In this animation we have only two keyframes: one in the beginning and one at the end. This means that the Composer (and ultimately the PME on the handheld device) is animating the object from the beginning to the end of the animation. If we want to have more control over it, we could add more keyframes. Let us add a keyframe at frame 25. Click frame 25, click the Add Key button, and click the Translation line. Grab the sun and move it to somewhere on the upper right of the canvas. Now your document window is showing the sun's new path in the sky, as shown in Figure 14-20.

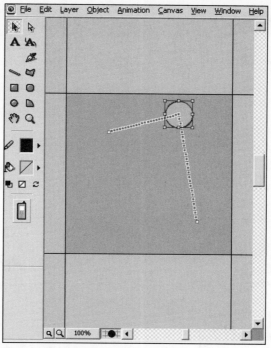

Figure 14-20: New sun path

If you select Play, you will see the sun animate along its new path. Since we did not change the color of the object at frame 25, Composer is still transitioning the color of the object from keyframe 1 to keyframe 50.

Tweening

You have already seen how Composer automatically calculates what to do to an object between keyframes. Now, what if you want to override these automatic transitions? *Tweening* enables you to do this. For example, let us modify the way our sun moves from keyframe 1 to keyframe 25. To do this, click the word Translation. The Tweening window pops up, as shown in Figure 14-21.

Change the path to Curving by using the drop-down selection in the Path field. You will now see that the object will follow a curved path between keyframe 1 and keyframe 25. Click OK and then ensure that frame 1 is selected. Select Play to see the animation.

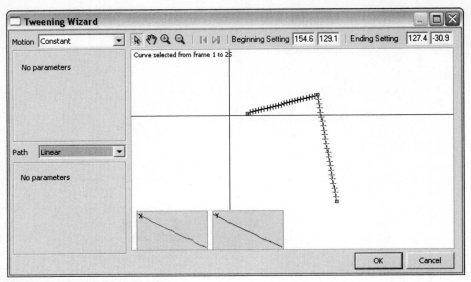

Figure 14-21: Tweening window

Event-Based Animation

Event-based animation enables you to take your animation to the next level. Not only can you create wonderful animations with Plazmic, but you can allow BlackBerry users to interact with the animation. This allows them to use the scroll wheel to move around the screen, play sounds, and link to other mobile media content. You can also use this interactivity to create *rollover objects*, which are objects that change when the mouse rolls over them. On a BlackBerry, these objects change when they are rolled over by using the scroll wheel.

Interactive Items

An interactive item comprises an *action* and an *event trigger*. The way in which you combine actions and event triggers enables you to create a variety of interactive effects.

The following table describes available actions.

Action	Description
Load Scene	Loads new content replacing the currently displaying content.
Play Animation	Play an animation sequence.
Play Sound	Plays a sound (typically a MIDI file).
Stop all Sounds	Stops playing all currently playing sound files.

Event triggers are user events that you capture and to which you assign an action. The following table describes available event triggers.

Event Trigger	Description
On Hotspot Focus-In	Specify what to do when a user scrolls onto a hotspot.
On Hotspot Focus-Out	Specify what to do when a user scrolls off a hotspot.
On Hotspot Activate	Specify what to do when a user activates a hotspot (clicks the scroll wheel or presses the space bar).

Creating Interactive Content

You can make the rising sun animation more interactive by adding a Start and Stop button. Each button will animate when it is rolled over. When the Start button is pressed, the sun will animate.

To begin, rename Layer_1 to Sun_Layer to make things easier to follow. Next, create a new layer and call it Stop_Layer. Create one more layer and call it Start_Layer. Select the Stop Layer and create a new text object. You can use the grid lines and Snap to Grid to help you line up the text. Place the object on the canvas, type **Stop**, and press Enter. Using the Inspector window, change the font to Arial and ensure that the vertical and horizontal sizes are both 13. Rename the object to Stop_Text. Next, select Start_Layer and repeat the steps for Stop_Layer, but this time type the word **Start**. Rename this object to Start_Text.

Next, you must create animation for your buttons. Select the Stop_Layer object, click the Stop_Text object, and then click frame 4 in the Animation window. Using the Inspector window, change the vertical and horizontal size of the text to 18. Now, click frame 8 of the animation and change the vertical and horizontal values to 13. Repeat the steps for the Start_Layer object with the Start_Text object.

At this point, you have created two new objects that can be animated. Because the Stop_Layer and Start_Layer objects are ahead of the Sun_Layer object, the Sun_Layer object will not animate when the document is viewed on a BlackBerry. In addition, because the Start_Layer object and Stop_Layer object are using interactivity, they will not animate either. Any animation will be driven by user events.

The next step is to set up how the user will interact with your new document. What you want to do is smoothly enlarge the text of each button when the user scrolls over it and smoothly reduce it when the user scrolls off it. If the user scrolls onto the Start button and clicks the scroll wheel, we want the sun to rise. To achieve this you must use the existing frames for your Start_Text and Stop_Text objects.

In the Layers window, right-click any layer and choose Interactivity. On the top left of this new window, shown in Figure 14-22, you will see your layers. On the bottom left, you will see your hotspots, and the order in which they will display on the screen. You can use the three buttons in the top right to filter the events displayed on the screen.

Simplify the screen by working with only certain kinds of user events at a time. Start with the On Focus In event. Click the On Focus In button. This will limit the display to only show On Focus In events. Click the Start_Layer object and then on the Create Event button (which is right next to the trash can icon), and choose Play Animation from the drop-down list. When you choose Play Animation, you are setting the action that occurs when the event is triggered. You must control that action by specifying which layer is animated and which frames of that animation are animated. You can set the animation to loop for a certain number of times, or indefinitely.

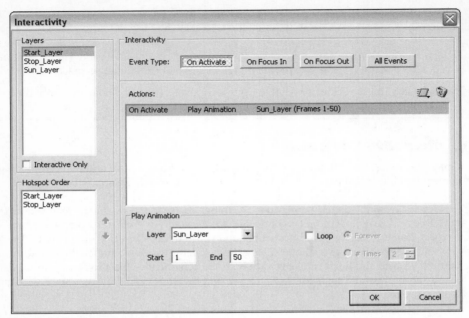

Figure 14-22: Interactivity window

Simplify the screen by working with only certain kinds of user events at a time. Start with the On Focus In event. Click the On Focus In button. This will limit the display to only show On Focus In events. Click the Start_Layer object and then on the Create Event button (which is right next to the trash can icon), and choose Play Animation from the drop-down list. When you choose Play Animation, you are setting the action that occurs when the event is triggered. You must control that action by specifying which layer is animated and which frames of that animation are animated. You can set the animation to loop for a certain number of times, or indefinitely.

For this example, you want to animate the Start_Layer and only from frame 1 to frame 4. By choosing these frames, you are displaying an animation of the Start_Text object enlarging. We want to do the same thing with the Stop_Layer. Click the Stop_Layer on the left of the screen, create a new event, and choose Play Animation. This time, choose to animate the Stop_Layer frames 1 to 4.

Now, you must show the text reducing in size when the user scrolls off the hotspot. To do this, click the On Focus Out button to limit the view to only those events. Click the Start_Layer object, click the Create Event icon, and choose Play Animation. This time, choose to animate the Start_Layer and frames 4 to 8. Click the Stop_Layer object and again create a Play Animation event. This time, animate the Stop_Layer and animate frames 4 to 8.

The last event and action must be the user's clicking (or activating) the Start button (or hotspot). Click the On Activate button. Click the Start_Layer object and then click the Create Event button. Once again choose Play Animation. This time, however, you will animate the Sun_Layer object and frames 1 to 50. In this example, the Stop button has no On Activate action because you cannot stop an animation in progress, so the button is just there for show.

To save the interactivity settings, click the OK button.

To test your animation, launch it in the simulator. Click the Simulator button in the Toolbox window. When the simulator loads, it will show the sun and the Start and Stop buttons. The simulator in the Plazmic CDK does not work in the same way as the simulator in the JDE. This means that the mouse scroll wheel does not move the simulator scroll wheel. To control the Plazmic BlackBerry simulator, you must move your mouse above the scroll wheel. While your mouse is there, an up arrow appears, as shown (circled) in Figure 14-23. To simulate moving the scroll wheel up, click the up arrow.

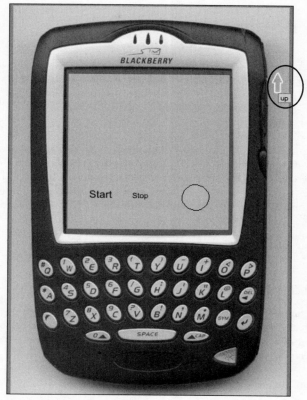

Figure 14-23: Plazmic simulator

The Simulator included with Composer is designed for developers and, therefore, is not an exact BlackBerry simulator. This simulator provides functions to pause, speed up, and restart animations. There is another BlackBerry simulator included in the Plazmic CDK, which we will cover later in this chapter in the section "Viewing Your Project."

To simulate clicking the scroll wheel, move the mouse to the right side of the scroll wheel and a left arrow will appear. If you click on the left arrow, the scroll wheel will click in.

In the simulation, you will see that as you move onto either the Start or Stop hotspot, the text will enlarge smoothly. As you scroll off the hotspot, the text will reduce smoothly. While on the Start hotspot, click the scroll wheel and the sun will animate. To exit the simulator, select File ⇨ Exit.

Timeline Events

In addition to user event triggers you can create *timeline events*, which trigger off of a certain frame in the animation. To create a timeline event, click the frame number in the Animation window, and then click the Timeline Actions button (it is the last button on the right next to the key buttons). When the Timeline Events window appears, you can create new actions. The types of actions available are exactly the same as the actions available in the Interactivity screen.

Exporting Your Document

After you have created your project and you want to make it available to your BlackBerry users, start by exporting it from the Composer. To export the files into a format that can be viewed immediately on a BlackBerry, select either PME or PMB formats. When you export the project to PME format, any bitmap images you used in the project are saved as separate files. To improve loading times of the Plazmic content over low-bandwidth networks, you can save the files in PMB format. The PMB file contains the animation and the bitmap images in one file.

You can also export your files to SVG format, which means that they can be used in other animation programs after being imported. If you have SVG files on your hard drive, they can be compiled later into either PME or PMB files by using the supplied `SVGC.EXE` command-line tool. The tool can be found in `c:\program files\plazmic cdk 3.7\bin`. When exporting or importing SVG files, it is important to remember that that Composer supports only a subset of the full SVG specification.

You can export your animation to animated GIF files. This option is useful only if you have no event triggers, because you cannot interact with an animated GIF.

Publishing Plazmic Content

To publish Plazmic content on the Web, add the relevant Multipurpose Internet Mail Extension (MIME) types to your Web server's configuration. The MIME types enable the browser to know what application to use when viewing content with a specific extension. Plazmic content can have the PME or PMB extension. The MIME types that need to be added are as follows:

❑ Type: **application/x-vnd.rim.pme**

  Extension: **pme**

❑ Type: **application/x-vnd.rim.pme.b**

  Extension: **pmb**

To add the correct MIME types to the Abyss Web server, open the Abyss Console and click Server Configuration. Then click the Advanced button. On the Advanced screen, click MIME Types. Scroll down to the bottom of the screen and click the Add button. On the next screen, type **application/x-vnd.rim.pme** into the Type field and **pme** into the Associated Extension field. Click OK to save the changes.

Next, click Add and type **application/x-vnd.rim.pme.b** in the Type field and **pmb** into the Associated Extension field. Click OK to save the changes. Restart the Abyss server before your changes will take effect.

If you are using another Web server (such as Apache or Microsoft IIS), consult the manufacturer's documentation to see how to add these MIME types to its configuration.

Exporting Your Project

You need to export your project so that it can be displayed in a browser. To do this, open the project in Composer. You have one that is pre-built called `anim1.cp`. This Composer document contains all of the steps you went through in the previous pages. Because we instructed you to use a font that does not reside on the BlackBerry, when you export the project bitmap, files will be exported for the font animation. To make the Plazmic file quicker to load in the browser, it is better to bundle the animation and bitmaps into one file. To do this, export the document as a PMB file. Your project will save as a 9 KB file. Copy the file to the root of your Web server.

Viewing Your Project

To view your project in the real world, you will need to place the PME file on the Web server. Next, start the MDS Simulator and then the BlackBerry Simulator. Use the simulator to browse to the particular file. Do not use the simulator in the Plazmic Composer program, but rather the one that is installed along with the Plazmic tools. You will find it in `start/programs/plazmic cdk 3.7 for blackberry/ blackberry handheld simulator/<blackberry type>`. The MDS Simulator is in `start/programs/ plazmic cdk 3.7 for blackberry/blackberry mobile data service simulator/`.

If you have installed the BlackBerry JDE, you can use that copy of the MDS and BlackBerry Simulator. Open the BlackBerry Browser, click the scroll wheel, and choose Go To. Enter **http://localhost/anim1.pmb**. If you have placed the file on a real Web server, replace *localhost* with your server name. Because you created this animation for the 7700 series BlackBerry with the larger screen, first view the file on that simulator.

When the animation loads, you can use the scroll wheel to move between the hotspots. Click the space bar to activate the hotspot. If you click the scroll wheel to activate the hotspot, you will somewhat spoil the effect because a menu pops up first and you have to choose Get Link.

When designing your Web portal or animations, you must be aware of the different screen sizes and attributes. As we discussed in Chapter 10, there is a way to first interrogate the handheld unit to find out its model number and software revision level. Once that has been obtained, the Web server can redirect the handheld unit to the correct content.

The other approach is to design your animation for one screen size. To get a sense of how this animation will look on a smaller screen, close the 7700 series simulator and load the 7200 series simulator. Repeat the steps to load the animation. When it is loaded, you will be presented with a blue screen. This is because you are looking at the top half of the screen. The hotspots and sun are in the lower part. To scroll down, click the scroll wheel and choose Scroll, as shown in Figure 14-24. Scroll down the screen until you reach the bottom, then click the scroll wheel again, and choose Stop Scrolling. Now the animation will work as before. Use the scroll wheel to move between the Start and Stop hotspots and press the spacebar to start the sunrise.

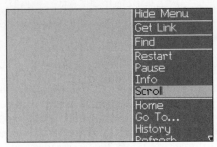

Figure 14-24: Scrolling down

If you close the 7200 series simulator and open the 6200 series, you will be able to see how your animation looks on a monochrome device. It does not look very attractive because it has not been designed for a monochrome device. However, if you want to, you can design animation for these handheld devices. Because you only have two colors to play with, you will not be able to do anti-aliasing or any kind of color changes in your animations. Other than that, you should be able to animate in the same way as you do with animations designed for color devices.

Summary

The Plazmic technology is a great way to create very attractive Web sites with full animation and vibrant colors. There really is no downside to this because the animation compresses to tiny files that are quick to load over low-bandwidth networks.

In this chapter, we have covered the main aspects of creating Plazmic content and we have used simple animation to showcase these aspects. When you create your Plazmic content, you will be able to use multiple combinations of user events and actions to create much more complex animations that will greatly enhance your user's browsing experience. To see examples of creative uses of Plazmic, load some of the sample files that were installed with the Plazmic CDK. In addition, visit `http://mobile` `.blackberry.com` to see RIM's showcase mobile site.

Wireless Markup Language Reference

The following table summarizes support for each Wireless Markup Language (WML) element and attribute (presented in alphabetical order). The browser supports WML 1.3.

Element	Description	Element Attributes	Description
<a>	Specifies a link to follow.	href	Specifies the target of the link. The browser substitutes any variable references from the Wireless Application Protocol (WAP) context.
		title	Ignored.
		accesskey	Ignored.
<access>	If specified, the browser compares the <access> element to the domain and path that are specified in the <access> element to determine whether the user has the proper access to display the page. If not, the browser displays an error message.	domain	Specifies the domain used to verify access privileges.

Table continued on following page

Element	Description	Element Attributes	Description
		path	Specifies the path used to verify access privileges.
<anchor>	When the user moves the cursor over a character or image that is contained in the <anchor> element, users can click the Follow Link menu item to perform the <go>, <prev>, or <refresh> task that is associated with the <anchor> element.	title	Ignored.
		accesskey	Ignored.
<b>	Renders text in bold in the current font and size, if available. Otherwise, the standard font is used.		
<big>	Renders text in the next larger size of the current font, if available. Nesting of <big> and <small> elements is respected.		
 	Starts a new line.		
<card>	Contains the entire card.	title	Displays the card title in the Title area of the screen.
		newcontext	If set to True, this attribute clears the variable in store.
		ordered	Ignored. All elements in a page are rendered.
		onenterforward	Opens the specified URL when the card is accessed using a <go> task.
		onenterbackward	Opens the specified URL when the card is accessed using a <prev> task.
		ontimer	Opens the specified URL when the timer has expired.

Element	Description	Element Attributes	Description
<do>	Defines an event trigger. The <do> element can enclose the following task elements: <go>, <prev>, <noop>, or <refresh>. In the browser, <do> elements that are defined on a card appear in two ways: as menu items on the browser menu or as soft keys in a non-scrolling area at the bottom of the screen.	type	Required. Specifies the type of <do> element. The <do> element of type can be one of accept, help, or prev. A <do> element of type appears first on the menu. The type attribute also matches <do> elements in the deck template, when the name attribute is not defined, for shadowing purposes.
		label	Specifies the label used for the associated menu item. If a label attribute is not included, a default name is assigned to the menu item based on type.
		name	Matches <do> elements in the deck template for shadowing purposes.
		optional	<do> elements with the optional attribute set appear after other <do> elements on the menu.
<em>	Renders text in italic in the current font, if available. Otherwise, the standard font is used.		
<fieldset>	Ignored. Enclosed elements are rendered as though the <fieldset> element did not exist.		

Table continued on following page

Element	Description	Element Attributes	Description
<go>	Directs the browser to a specified URI. The browser sets any variables that are specified in <setvar> elements within the <go> element, and includes any values that are specified in contained <postfield> elements.	href	Specifies the target URI that the browser goes to.
		sendreferer	Specifies whether the URI is sent in the request. Valid values are yes or no.
		method	Specifies the request type. The browser supports both the GET and the POST methods of sending requests.
		enctype	Ignored. The browser uses the default: application/x-www-form-urlencoded.
		cache-control	Forces page retrieval from the network instead of the cache. If this attribute is set to no-cache, a flag is set on the request being sent to invalidate this page in the cache (if it is there), before retrieving the page.
		accept-charset	Ignored. The browser uses the default: UTF-8.
<head>	Ignored. See the <access> and <meta> elements.		
<i>	Renders text in italic in the current font and size, if available. Otherwise, the standard font and size is used.		
<img>	Defines an image. When images are contained in <anchor> or <a> elements, users can select the image.	alt	Specifies the text that appears when an image is unavailable, provided that the browser has been correctly configured to display it. When the real image arrives, the real image replaces the alt placeholder.

Element	Description	Element Attributes	Description
		src	Specifies the location of the image on the server.
		localsrc	Unsupported. The browser does not support <localsrc> images.
		vspace	Specifies the amount of white space to insert above and below the image.
		hspace	Specifies the amount of white space to insert to the left and right of the image.
		align	Specifies the horizontal alignment of the image on the line.
		height	Specifies the image height.
		width	Specifies the image width.
<input>	Renders as a text field into which users can type text. If users type a value that is not consistent with the restrictions specified by the attributes in the input element, the browser displays a warning when users try to save the value. Input boxes appear on a separate line from the surrounding text.	name	Specifies the name of the variable to set with any typed value.
		type	Specifies the type of information being collected. Valid values include text or password. If type is set to password, the browser displays an asterisk (*) for each character that the user types.
		value	Specifies the initial value that is displayed in the input field when the card is first rendered. This value is used only when the variable specified by the name attribute is not set already in the WAP context. If the variable is set, the current value of that variable is used instead.

Table continued on following page

Element	Description	Element Attributes	Description
		format	Specifies a mask. The browser checks that any text the user enters in the input field conforms to the mask that is specified by this attribute. The following character tags define which characters can be typed: A for any symbolic or uppercase alphabetic character (no numbers); a for any symbol or lowercase alphabetic character (no numbers); N for any numeric character (no symbols or alphabetic characters); X for any symbolic, numeric, or uppercase alphabetic character (not changeable to lowercase); x for any symbolic, numeric, or lowercase alphabetic character (not changeable to uppercase); M (the default) for any symbolic, numeric, or lowercase alphabetic character (changeable to lowercase) (Note: for multiple character input, the user input automatically defaults to an uppercase first character); m for any symbolic, numeric, or lowercase alphabetic character (changeable to uppercase) (Note: for multiple character input, the user input automatically defaults to a lowercase first character). **Tips:** To limit the number of characters users can type, specify a single-digit number before the character tag. For example, 3X requires the user to type a maximum of three symbolic, numeric, or uppercase alphabetic characters. To enable users to type an unlimited number of characters, specify an asterisk (*) before the character tag. For example, *a enables the user to type any number of symbolic or lowercase alphabetic characters. To insert a character into the mask, use the syntax \c, replacing the c with the character that you want to insert. This is useful for inserting, for example, a dash in a nine-digit area code.
		emptyok	Enables the user to leave the field blank.
		size	Specifies the size (in characters) of the input box.
		maxlength	Specifies the maximum length of text that the user can type.
		tabindex	Ignored. Tabbing is not supported.

Element	Description	Element Attributes	Description
`<meta>`	Specifies generic name-value pairs relating to the deck.	name	Identifies the meta-information that is defined. The `<meta>` element is ignored if the name attribute is not specified.
		http-equiv	Specifies whether the content of the `<meta>` element is bound to an HTTP response header. The browser examines the `<meta>` element to determine whether it is relevant and processes it accordingly.
		forua	Specifies whether the metadata is sent to the browser. If set to True, the metadata is sent to the browser. This attribute is ignored if it is set to False.
		content	Specifies the HTTP header value.
		scheme	Ignored.
`<noop>`	Specifies whether the browser effectively removes any `<do>` or `<onevent>` elements from the current card.		
`<onevent>`	Specifies how events are handled.	type	Required. Identifies the event to handle; it also matches `<do>` elements in the deck template for shadowing. The type attribute can have one of the following values: onenterbackward (occurs when a user navigates backward to a card); onenterforward (occurs when a user navigates forward to a card); onpick (occurs when a user selects or clears an option); ontimer (occurs when the timer, specified by the `<timer>` element, expires).
		id	Sets a unique name for the element.
`<optgroup>`	Ignored. Options in a `<select>` element are rendered as a flat list on the screen.		

Table continued on following page

Element	Description	Element Attributes	Description
\<option\>	Specifies an item in a \<select\> list. \<option\> elements are rendered in the manner specified by the enclosing \<select\> list.	value	Specifies the value of the enclosing \<select\> variable, if the \<select\> element has a NAME attribute.
		title	Ignored.
		onpick	Specifies the URI that the browser loads when the option is selected.
\<p\>	Denotes a new paragraph and renders content according to the attributes. For layout purposes, the \<p\> and \<br/\> elements are the same.	align	Specifies the position of the contents of the \<p\> element on screen.
		mode	Ignored. Content is always wrapped.
\<postfield\>	Specifies name/value pairs that are included in the HTTP request. \<postfield\> elements must be enclosed in a \<go\> element. Variable references in the name and value attributes of each \<postfield\> element are replaced with appropriate values from the WAP context.	name	Required. Specifies the name of the variable to be passed to the server.
		value	Required. Specifies the value of the variable to be passed to the server.
		id	Sets a unique name for the element.
\<pre\>	Renders text using a fixed-space font.		

Element	Description	Element Attributes	Description
<prev>	Directs the browser to a specified URI. The inclusion of the <prev> element in <do> elements affects menu construction. The browser menu always contains at least one item that enables users to go back in the navigation history. If the current card does not contain any <do> elements with a <prev> task, by default the browser creates a Back menu item. If the current card contains one or more <do> elements that contain <prev> elements, the browser does not create a separate menu item and relies on the <do> elements for that card to provide this behavior instead.		
<refresh>	Refreshes any variables that are specified by the enclosed <setvar> element(s). If an event occurs that has a <refresh> task, the browser first sets any variables that are specified in <setvar> elements that are contained in the <refresh> element. It then refreshes the current page using the updated WAP context.		

Table continued on following page

Element	Description	Element Attributes	Description
<select>	Presents a list to the user. Users can select one or more of the options. Select lists are fully displayed on every card. Users can scroll through options. Single-selection lists are rendered as options buttons. Multiple-selection lists are rendered as check boxes.	title	Ignored.
		name	Specifies the name of the variable that is set with the selection.
		value	Sets the default value of the variable.
		iname	Specifies the variable to be set with the (1-based) index of the selected option, according to the specifications.
		ivalue	Sets the index(es) of the preselected option(s). Use this attribute only if the variable specified by iname does not already have a value.
		multiple	Specifies whether the <select> element is rendered as a multiple-selection list with a set of check boxes.
		tabindex	Ignored. Tabs are not supported.
<setvar>	Specifies a new value for a given variable in the WAP context. When a <go>, <prev>, or <refresh> task is performed, the browser first looks for any associated <setvar> elements and then updates the WAP context accordingly before running the task.	name	Sets the name of the variable.
		value	Sets the value of the variable.

Element	Description	Element Attributes	Description
<small>	Renders text in the smaller size of the current font, if available. The browser supports nesting of big and small elements.		
<strong>	Renders text in the bold font of the current size, if available. Otherwise, the standard font is used.		
<table>	Denotes a new table.	title	Ignored.
		align	Aligns the text table columns according to the align value. This attribute is optional. The content of each cell is left-aligned by default. For example, with align="LCR", content in the first column is left-justified; content in the second column is centered; and content in the third column is right-aligned.
		columns	Required. Specifies the number of columns in the table.
		tr	Denotes a new table row.
		td	Denotes a new table cell.
<template>	Specifies deck-level <do> or <onevent> items. These items are included in every card for the deck, unless the card has a more specific <do> or <onevent> element that shadows those in the template. The browser handles template attributes as though they are <onevent> definitions for the corresponding events.	onenterforward	Treated as an onenterforward <onevent> element with a <go> action.
		onenterbackward	Treated as an onenterbackward <onevent> element with a <go> action.

Table continued on following page

227

Element	Description	Element Attributes	Description
		ontimer	Treated as an ontimer <onevent> element with a <go> action.
<timer>	Specifies a timer for the current card. The browser implements card timers according to the WML specifications. In particular, a <refresh> operation stops the timer, sets its corresponding variable to the current timer value, performs the refresh, and then resets the timer and restarts it. When the timer expires, the variable that is specified by the name attribute is set to 0 before any <ontimer> tasks are run.	name	Specifies the variable name to be set with the timer value for card entry, exit, and timer-expire events.
		value	Specifies the initial timer value for on-card entry events.
<u>	Underlines content enclosed in the <u> element in the current font and size, if available. Otherwise, the standard font and size are used.		

B

WMLScript Compendium

This article by Richard Evers (Editor) appeared in the Blackberry Developer Journal *1, no. 3 (October 2004) and is reprinted by permission subject to the legal disclaimer set forth at* www.blackberry.com/ developers/journal/jan_2005/legal_disclaimer.shtml. *Research In Motion Limited (RIM) assumes no responsibility for any typographical, technical, or other inaccuracies in this document. No representations or warranties by RIM may be inferred from this document, and RIM shall not be liable for any damages or other harm attributable directly or indirectly to the information contained herein.*

A new plague hit the world in 1995 in the form of hack generalists who could copy and paste together awful HTML sites.

As the years trickled by, and dot com bubbles burst, bandwidth improved along with the languages, techniques and software used to create really decent web sites.

While many conventional sites are pretty good now, the wireless world of WAP is still wallowing in an awkward stage of existence. Unlike traditional sites, WAP sites look basic at best because they are constrained to work within the confines of wireless devices with small screens, modest memory and limited bandwidth.

The Wireless Markup Language (WML) is a broad, fairly easy to use language that permits site developers to create reasonable content that can be viewed on all WAP browsers. Unfortunately, it lacks many of the useful things found in a true scripting language.

The solution is WMLScript, a wireless scripting language that can co-exist with WML decks. It is similar to JavaScript, and is a modified subset of ECMAScript.

WMLScripts are independent files called as external references within WML decks. They are compiled into byte code at run time on the server before being sent to the WAP browser. Like C, C++ and Java, the language is case sensitive and has a construct that is similar to C.

The format of a WMLScript function is as follows:

```
extern function NAME( [PARAMETERS(S)] )
{
  // body of function
}
```

The `extern` keyword is used to make the function public to external files. Do not include `extern` on functions that are only called from within the script.

The following example consists of a WML deck that calls an external WMLScript function:

```
<?xml version="1.0"?>
<!DOCTYPE wml PUBLIC "-//WAPFORUM//DTD
  WML 1.1//EN" "http://www.wapforum.org/DTD/wml_1.1.xml">
<wml>
  <card id="page1" title="Execute Script">
    <do type="options" label="GetNews">
      <go href="go_mobile.wmls#surf('news')"/>
    </do>
    <do type="options" label="GetFlightStat">
      <go href="go_mobile.wmls#surf('flightstat')"/>
    </do>
  </card>
</wml>
```

The highlighted lines in the earlier example contain references to the external WMLScript to follow:

```
/*
 * Filename: go_mobile.wmls
 * Function: surf()
 * Purpose : selectively surf the mobile web
 */
extern function surf(the_url)
{
  if (the_url == "news")
  {
      WMLBrowser.go("http://mobile.globeandmail.com");
  }
  else if (the_url == "flightstat")
  {
    WMLBrowser.go("http://mobile.aircanada.ca/aircanada/flstatus.wml");
  }
}
```

Parameter passing is fairly lax where type is not specified in the parameter list. The caller of the function is responsible for making sure that parameters passed to a function are in the expected format and sequence.

Reserved Words

access	domain	http	struct
agent	else	if	super
break	equiv	import	switch
catch	enum	isvalid	throw
class	export	meta	Try
const	extends	name	typeof
continue	extern	path	url
debugger	finally	private	use
default	for	public	user
div	function	return	var
do	header	sizeof	while

Reserved words cannot be used to name functions or variables.

Statements

WMLScript statements are as follows:

break
continue
for
if/else
return
while

Operators and Expressions

Arithmetic operators

+	Plus
*	Multiply
%	Remainder

Table continued on following page

-	Minus
/	Divide
div	Perform integer division

Bitwise Operators

<<	Left shift
&	AND
^	Exclusive OR
>>	Right shift
|	OR
>>>	Bitwise right shift with zero fill

Assignment Operators

=	Assignment
-=	Subtract and assign
/=	Divide and assign
%=	Remainder and assign
>>=	Bitwise right shift and assign
&=	Bitwise AND and assign
^=	Bitwise XOR and assign
+=	Add and assign
*=	Multiply and assign
div=	Integer divide and assign
<<=	Bitwise left shift and assign
>>>=	Bitwise right shift, zero fill and assign
|=	Bitwise OR and assign

Unary Operators

+	Plus
--	Post or pre-decrement

~	Bitwise NOT
-	Minus
++	Post or pre-increment

Logical Operators

&&	AND
!	NOT
||	OR

String Operators

| + | Concatenate |
| += | Concatenate |

Comparison Operators

<	Less than
==	Equal
>=	Greater than or equal
<=	Less than or equal
>	Greater than
!=	Not equal

Conditional Operator

? :

Example:

```
var IsOkay = ( Want == "Food" ) ? 1 : 0;
```

Other operators

WLMScript supports comma-operators:

```
for (el = 0, id = 100; el < 10; el++, id+=10)
   // do something
```

While WMLScript is weakly typed, it does support Boolean, integer, floating-point, string, and invalid data types. The `typeof` operator will return an integer value that identifies the data type, as shown below:

1	integer
2	floating-point
3	string
4	boolean
5	invalid

Example:

```
var data = "Mares eat oats";
var result = typeof data; // result equals 2
```

An `isvalid` operator is provided to safely test whether an expression is valid. It will return `true` if the passed expression is valid, else returns `false`. The syntax is as follows:

```
var IsOk_1 = isvalid (99/0); // false
var IsOk_2 = isvalid (99/1); // true
```

Libraries

The strength of WMLScript is largely contained with the function libraries. The libraries include functions to deal with numeric values, dialog and alerts, strings, relative and absolute URLs, and the browser.

Library: Lang

Lang contains the core library functions:

```
abort

abs

characterSet

exit

isInt

max

maxInt

min

minInt

parseInt
```

```
random

seed
```

Lang.abort()

Description:

Aborts execution of WMLScript and returns passed string to caller.

Syntax:

```
Lang.abort(ErrorMessage);
```

Parameters:

An error message.

Returns:

Does not return.

Example:

```
Lang.abort("Script failure on line 123");
```

Lang.abs()

Description:

Returns the absolute value of the passed number.

Syntax:

```
Value = Lang.abs(Number);
```

Parameters:

An integer or float value.

Returns:

The absolute value returned as an integer or float value.

Example:

```
// will return positive 98765
var i_ret = Lang.abs(-98765);

// will return positive 987.65
var f_ret = Lang.abs(-987.65);
```

Lang.characterSet()

Description:

Returns a value that identifies the supported character set.

> Visit **http://www.iana.org/assignments/character-sets** *for an up-to-date list of character sets.*

Syntax:

```
charset = Lang.characterSet();
```

Parameters:

```
void
```

Returns:

Numeric character-set identifier.

Example:

```
// Assigned MIB enum Numbers
//------------------------
//0-2        Reserved
//3-999      Set By Standards Organizations
//1000-1999  Unicode and ISO/IEC 10646
//2000-2999  Vendor
// for example, returns 3 for US-ASCII
var charset = Lang.characterSet();
```

Lang.exit()

Description:

Terminates script and returns passed message to caller.

Syntax:

```
Lang.exit(Message);
```

Parameters:

A message to pass back to the caller.

Returns:

Does not return.

Example:

```
Lang.exit("Exit stage left");
```

Lang.isInt()

Description:

Tests if the passed string can be converted into an integer value using `parseInt()`.

Syntax:

```
isOkay = Lang.isInt(StringValue);
```

Parameters:

A string representation of an integer value.

Returns:

Boolean `True` if string will convert to integer form; else `False`.

Example:

```
isOkay1 = Lang.isInt("98765"); // true
isOkay2 = Lang.isInt("-98765"); // true
isOkay3 = Lang.isInt("9.8e2"); // true
isOkay4 = Lang.isInt("intni"); // false
```

Lang.max()

Description:

Determines the maximum value of two passed values in either integer or floating-point form.

Syntax:

```
maxValue = Lang.max(Value1, Value2);
```

Parameters:

Two integer or floating-point values to compare.

Returns:

The highest value passed.

Example:

```
// returns 13
var maxValue = Lang.max(12, 13);
```

Lang.maxInt()

Description:

Returns the maximum value of an integer.

Syntax:

```
theMax = Lang.maxInt();
```

Parameters:

```
void
```

Returns:

Maximum integer value.

Example:

```
// returns 2147483647
var theMax = Lang.maxInt();
```

Lang.min()

Description:

Returns the minimum value of two passed values in either integer or floating-point form.

Syntax:

```
theMin = Lang.max(Value1, Value2);
```

Parameters:

Two integer or floating-point values to compare.

Returns:

The smallest value passed.

Example:

```
// returns 12
var theMin = Lang.max(12, 13);
```

Lang.minInt()

Description:

Returns the minimum value of an integer.

Syntax:

```
theMin = Lang.minInt();
```

Parameters:

```
void
```

Returns:

Minimum integer value.

Example:

```
// returns -2147483648
var theMin = Lang.maxInt();
```

Lang.parseInt()

Description:

Converts a string into an integer value.

Syntax:

```
theInt = Lang.parseInt(StringInt);
```

Parameters:

Integer value in string form.

Returns:

Integer value.

Example:

```
// returns 9876
var theInt1 = Lang.parseInt("9876");

// returns 9876
// stops parsing on first error
var theInt2 = Lang.parseInt("9876Hi!!");
```

Lang.random()

Description:

Returns a random number between 0 and the passed value.

Syntax:

```
rndValue = Land.random(MaxRange);
```

Parameters:

Maximum integer value to draw from.

Returns:

A random integer within the specified range.

Example:

```
var rndValue = Land.random(9867);
```

Lang.seed()

Description:

Initializes the random number generator.

Syntax:

```
ret = Lang.seed(SeedValue)
```

Parameters:

An integer seed value.

Returns:

An empty string.

Example:

```
var ret = Lang.seed(98765)
```

Library: Dialogs

`Dialogs` contains three dialog handlers:

alert

confirm

prompt

Dialogs.alert()

Description:

Displays passed message and waits for a response.

Syntax:

```
var ret = Dialogs.alert(Message);
```

Parameters:

Message to display.

Returns:

An empty string.

Example:

```
var ret = Dialogs.alert("Wake up now!");
```

Dialogs.confirm()

Description:

Displays passed message, waits for a response, then returns a Boolean value that corresponds to the selected option.

Syntax:

```
var isOkay = Dialogs.confirm(Message, Okay, Cancel);
```

Parameters:

- ❏ All parameters are string or string literals
- ❏ #1: the confirmation message
- ❏ #2: user option #1/2
- ❏ #3: user option #2/2

Returns:

Boolean True if parameter 2 is selected; else, False.

Example:

```
var isOkay = Dialogs.confirm("Exit", "Yes", "No");
```

Dialogs.prompt()

Description:

Displays passed message, waits for a response, then returns the user's response, or the default response if nothing had been entered.

Syntax:

```
var gotIt = Dialogs.prompt(Message, Default);
```

Parameters:

- ❏ All parameters are string or string literals
- ❏ #1: the prompt message
- ❏ #2: default response if nothing entered

Returns:

String response.

Example:

```
var Age = Dialogs.prompt("Your age", "30");
```

Library: *String*

String contains 16 functions:

```
charAt

compare

elementAt

elements

find

format

insertAt

isEmpty

length

removeAt

replace

replaceAt

squeeze

subString

toString

trim
```

String.charAt()

Description:

Returns a single character located at the passed offset position within the passed string.

Syntax:

```
var theChar = String.charAt(Buffer,Offset);
```

Parameters:

- ❏ #1: source string buffer
- ❏ #2: offset position within source buffer

Returns:

A single character located at the offset position.

Example:

```
// return "c"
var theChar1 = String.charAt("abcdef",3);

// returns ""
var theChar2 = String.charAt("abcdef",8);
```

String.compare()

Description:

String comparison where ranking is performed based on the ASCII value of each character within the string.

Syntax:

```
var theRes = String.compare(String1, String2);
```

Parameters:

- ❏ #1: first string to compare
- ❏ #2: second string to compare

Returns:

- ❏ 0 if the strings are identical
- ❏ -1 if the first string is less than the second
- ❏ 1 if the second string is less than the first

Example:

```
// return 0
var theRes1 = String.compare("ABC", "ABC");

// returns 1
var theRes2 = String.compare("abc", "ABC");

// returns -1
var theRes3 = String.compare("ABC", "abc");
```

String.elementAt()

Description:

Locates a single element within passed string buffer.

Syntax:

```
var Field = String.elementAt(Buffer, Element, Delim);
```

Parameters:

- ❑ #1: string buffer containing delimited fields
- ❑ #2: numeric value set to the desired field number
- ❑ #3: character(s) used to delimit fields

Returns:

The requested field, or the first field if a negative element is passed, or the last field if the element exceeds the total number of fields.

Example:

```
// returns "transformation"
var Field = String.elementAt("In an extraordinary transformation of heat to light,
Gibbon rested.", 4, " ");
```

String.elements()

Description:

Counts how many times a delimiter occurs within a passed buffer.

Syntax:

```
var howMany = String.elements(Buffer, Delimiter);
```

Parameters:

- ❑ #1: string buffer
- ❑ #2: field delimiter

Returns:

The total count of Delimiter within the Buffer.

Example:

```
// returns 20
var howMany = String.elements("This descent from unity into multiplicity recalled
Constantine's timid policy of 'dividing whatever is united', but its effects were
far different", " ");
```

String.find()

Description:

Searches for the first occurrence of the passed substring with the passed buffer.

Syntax:

```
var theFirst = String.find( Buffer, SubString );
```

Parameters:

❑   #1: source string buffer

❑   #2: substring to search for within passed buffer

Returns:

Offset value of first occurrence of substring (0-n), or -1 if substring is not found.

Example:

```
// returns 2
var theFirst = String.find( "Waterloo", "ter" );
```

String.format()

Description:

Formats the passed numeric value as a string.

Syntax:

```
var looksNice = String.format( FormatString, Value );
```

Parameters:

#1: Formatting string configured as follows:

```
%[width][.precision] type
```

where % and type are mandatory

width	Minimum number of characters that must be returned in the string.
.precision	Required decimal precision that is set based on the setting of type.
type='d'	Source is treated as a positive or negative integer value in the form of [-]9999, where 9999 is one or more decimal digits. If .precision is set, then the output value is padded on the left side with up to the .precision number of zeroes.
type='f'	Source is treated as a positive or negative floating-point value in the form of [-]9999.9999, where 9999 is one or more decimal digits. If .precision is set, then is used to set the number of digits after the decimal point, with at least one digit appearing before the decimal point. The default precision is 6. If a 0 or nothing has been specified after the '.' then the decimal component is truncated.
type='s'	Source is treated as a string. The width argument can be used to set the minimum string size. The .precision argument can be used to set the maximum string size.

Returns:

A formatted string.

Example:

```
// returns "9876"
var s1 = String.format("%8d", 9876);

// returns "009876"
var s2 = String.format("%8.6d", 9876);

// returns "9876.543"
var s3 = String.format("%8.3f", 9876.54321);

// returns "NCC-1701"
var s4 = String.format("NCC-%4d", 1701);

// returns "Hobbits rule!"
var s5 = String.format("Hobbits %s", "rule!");

// returns "        98.765%"
var s6 = String.format("%10.3F%%", 98.7654);
```

String.insertAt()

Description:

Creates a new string from the passed buffer that includes the passed field and field delimiter, inserted at the passed field element number.

Syntax:

```
var nStr = String.insertAt(Buff, Field, Element, Delim);
```

Parameters:

❏   #1: string buffer containing delimited fields.

❏   #2: string field to insert.

❏   #3: Numeric element (0-n) where the field is to be inserted. If less than 0 then is 0 is used. If greater than maximum number of elements, then the new field is appended to the buffer.

❏   #4: Character delimited to insert after the Field.

Returns:

Resulting string.

Example:

```
// results: "1|99|2|"
var nStr = String.insertAt("1|2|","99",1,"|");
```

String.isEmpty()

Description:

Determines is the passed string is empty.

Syntax:

```
var IsNULL = String.isEmpty(Buffer);
```

Parameters:

A string buffer.

Returns:

Boolean `True` if the string is empty; else `False`.

Example:

```
// returns true
var NULLString = "";
var IsNULL = String.isEmpty(NULLString);
```

String.length()

Description:

Returns the length of a passed string.

Syntax:

```
var theLength = String.length(Buffer);
```

Parameters:

A string buffer.

Returns:

The string length (0-n).

Example:

```
// returns 6
var theLength = String.length("yellow");

// returns 0
var theLength = String.length("");
```

String.removeAt()

Description:

Removes a field from passed buffer at a specific element position.

Syntax:

```
var Res = String.removeAt(Buffer,Element,Delim);
```

Parameters:

- ❑ #1: source string buffer
- ❑ #2: field element to remove from buffer
- ❑ #3: character delimiter used to separate fields

Returns:

String buffer without the requested element.

Example:

```
// returns  "1|3|"
var Res = String.removeAt("1|2|3|",1,"|");
```

String.squeeze()

Description:

Creates a string where all repeat white spaces in the passed string buffer are reduced to single spaces.

Syntax:

```
var SqzMe = String.squeeze(Buffer);
```

Parameters:

A string buffer.

Returns:

The string buffer with "squeezed" spaces.

Example:

```
// return "Will B Good"
var SqzMe = String.squeeze("Will  B  Good");
```

String.subString()

Description:

Returns a portion of the passed string.

Syntax:

```
var SS = String.subString(Buffer, Start, Size);
```

Parameters:

- ❏   #1: source string buffer
- ❏   #2: starting offset into the source string (0-n)
- ❏   #3: the number of characters to extract

Returns:

The requested substring.

Example:

```
// returns "ffe"
var SS = String.subString("Coffee", 2, 3);
```

String.toString()

Description:

Returns a string representation of the passed parameter.

Syntax:

```
var theString = String.toString(theValue);
```

Parameters:

Anything.

Returns:

A string.

Example:

```
var theString = String.toString(98.76);
```

String.trim()

Description:

Returns passed string without leading and trailing spaces.

Syntax:

```
var isTrimmed = String.trim(Buffer);
```

Parameters:

A string buffer.

Returns:

The string buffer without leading and trailing spaces.

Example:

```
// returns "Well Padded"
var isTrimmed = String.trim(" Well Padded "
```

Library: URL

URL contains 14 functions:

```
escapeString

getBase

getFragment

getHost

getParameters

getPath

getPort

getQuery

getReferer

getScheme
```

```
isValid

loadString

resolve

unescapeString
```

URL.escapeString()

Description:

Returns a string where special characters are changed into hexadecimal escape sequences. The escaped characters are as follows:

- ❑ Control Characters (ASCII %00 to %1F) and %7F
- ❑ Space (ASCII %20)
- ❑ Upper range (ASCII %8F to %FF)

Reserved characters:

```
;

/

?

:

@

&

=

+

$
```

Not recommended characters:

```
{

}

|

\

^

[

]

`
```

Delimiters:

```
<
>
#
%
"
```

Syntax:

```
var newStr = URL.escapeString(theURL);
```

Parameters:

String buffer containing unescaped URL.

Returns:

String buffer containing escaped URL.

Example:

```
// results: "http%3a%2f%2frim.com%2f"
URL.escapeString("http://rim.com/");
```

URL.getBase()

Description:

Returns the absolute URL (without fragment) of the current WMLScript.

Syntax:

```
var absURL = URL.getBase();
```

Parameters:

```
void
```

Returns:

String of absolute URL.

Example:

```
// if URL = "http://rim.com/script.wmls#frag"
// then returns "http://rim.com/script.wmls"
var absURL = URL.getBase();
```

URL.getFragment()

Description:

Returns the fragment portion of the passed URL.

Syntax:

```
var theFrag = URL.getFragment(theURL)
```

Parameters:

A URL.

Returns:

URL fragment.

Example:

```
// returns "frag"
var theURL = "http://rim.com/script.wmls#frag"
var theFrag = URL.getFragment(theURL);
```

URL.getHost()

Description:

Returns the host specified within the passed URL.

Syntax:

```
var theHost = URL.getHost(theURL);
```

Parameters:

A URL.

Returns:

Host component.

Example:

```
// returns "www.rim.com"
var theURL = "http://www.rim.com/script.wmls";
var theHost = URL.getHost(theURL);

// returns ""
theURL = "script.wmls";
theHost = URL.getHost(theURL);
```

URL.getParameters()

Description:

Returns the parameters within the last path segment of the passed URL.

Syntax:

```
var parms = URL.getParameters(theURL);
```

Parameters:

A URL.

Returns:

The parameters.

Example:

```
// returns "foo;bar"
var theURL = "http://rim.com/foo.php;foo;bar";
var parms = URL.getParameters(theURL);

// returns ""
theURL = "http://www.rim.com/script.wmls";
parms = URL.getParameters(theURL);
```

URL.getPath()

Description:

Returns the path specified within the passed URL.

Syntax:

```
var thePath = URL.getPath(theURL);
```

Parameters:

A URL.

Returns:

The path component.

Example:

```
// returns "/foo/bar.php"
var theURL = "http://rim.com/foo/bar.php";
var thePath = URL.getPath(theURL);

// returns ""
theURL = "http://rim.com/";
thePath = URL.getPath(theURL);
```

URL.getPort()

Description:

Returns the port specified within the passed URL.

Syntax:

```
var thePort = URL.getPort(theURL);
```

Parameters:

A URL.

Returns:

The port component as a string.

Example:

```
// returns "80"
var theURL = "http://www.rim.com:80";
var thePort = URL.getPort(theURL);

// returns ""
theURL = "http://www.rim.com";
thePort = URL.getPort(theURL);
```

URL.getQuery()

Description:

Returns the query portion of the passed URL.

Syntax:

```
var theQ = URL.Query(theURL);
```

Parameters:

A URL.

Returns:

The query.

Example:

```
// returns "bar"
var theURL = "http://rim.com/ok.php?foo=bar";
var theQ = URL.getQuery(theURL);

// returns ""
thePort = URL.getPort("http://www.rim.com");
```

URL.getReferer()

Description:

Returns the smallest relative URL for the page, deck or script that called the current script.

Syntax:

```
var whoCalled = URL.getReferer();
```

Parameters:

```
void
```

Returns:

The referrer.

Example:

```
// might return full URL
// or something relative such as "mydeck.wml"
// it will return "" if there is no referer
var whoCalled = URL.getReferer();
```

URL.getScheme()

Description:

Returns the scheme within the passed URL.

Syntax:

```
var theScheme = URL.getScheme(theURL);
```

Parameters:

A URL.

Returns:

The scheme.

Example:

```
// returns "http"
var theURL = "http://www.rim.com/";
var theScheme = URL.getScheme(theURL);

// returns ""
var theURL = "www.rim.com/";
var theScheme = URL.getScheme(theURL);
```

URL.isValid()

Description:

Validates the syntax of the passed URL.

Syntax:

```
var isOkay = URL.isValid(theURL);
```

Parameters:

A URL.

Returns:

Boolean `True` if syntax is correct; else returns `False`.

Example:

```
// returns true
var theURL = "http://www.rim.com/";
var isOkay = URL.isValid(theURL);

// returns false
theURL = "http:/www.rim.com/";
isOkay = URL.isValid(theURL);
```

URL.loadString()

Description:

Returns the content referred by the passed absolute URL and content type.

Syntax:

```
var theContent = URL.loadString(theURL, theCT);
```

Parameters:

❑ #1: string containing an absolute

❑ #2: string containing the Content Type that must prefix with `"text/"`.

Returns:

A string buffer containing the requested page, deck or script.

Example:

```
// returns the page contents
var theURL = "http://www.rim.com/index.shtml";
var theCT  = "text/plain" );
var theContent = URL.loadString(theURL, theCT);
```

URL.resolve()

Description:

Combines the passed base and relative URLs to return an absolute URL.

Syntax:

```
var absURL = URL.resolve(baseURL, relURL);
```

Parameters:

❑   #1: base URL (for example, `"http://www.rim.com/"`)

❑   #2: relative URL (for example, `"index.shtml"`)

Returns:

The resulting absolute URL.

Example:

```
// returns "http://www.rim.com/index.shtml"
var baseURL = "http://www.rim.com/";
var relURL = "index.shtml" );
var absURL = URL.resolve(baseURL, relURL);
```

URL.unescapeString()

Description:

Returns a string where escaped characters have been restored to original form.

Syntax:

```
var newURL = URL.unescapeString(escURL);
```

Parameters:

An escaped URL.

Returns:

An unescaped URL.

Example:

```
// results: "http://rim.com/"
var escURL = "http%3a%2f%2frim.com%2f";
var newURL = URL.unescapeString(escURL);
```

Library: Browser

Browser contains seven functions:

```
getCurrentCard

getVar

go

newContext

prev

refresh

setVar
```

Browser.getCurrentCard()

Description:

Returns the smallest relative URL of the current card being processed by the browser. If the current card has a different base than the current script, will return the absolute URL of the card.

Syntax:

```
var relURL = Browser.getCurrentCard();
```

Parameters:

```
void
```

Returns:

A relative or absolute URL.

Example:

```
// results: e.g. "validate#doit"
var relURL = Browser.getCurrentCard();
```

Browser.getVar()

Description:

Returns the value of the passed variable name within the current browser context.

Syntax:

```
var varVal = Browser.getVar(strName);
```

Parameters:

Variable name.

Returns:

String value or invalid if not found.

Example:

```
// results: e.g. "jdoe"
var varVal = Browser.getVar("userid");
```

Browser.go()

Description:

Navigates browser to a URL.

Syntax:

```
var ret = Browser.go(navURL);
```

Parameters:

A relative or absolute URL.

Returns:

An empty string.

Example:

```
// relative navigation
var ret = Browser.go("newpage.wml");

// absolute navigation
var ret = Browser.go("http://www.xyzzy.com/");
```

Browser.newContext()

Description:

Resets browser context thus clearing all variables.

Syntax:

```
var ret = Browser.newContext();
```

Parameters:

```
void
```

Returns:

An empty string.

Example:

```
var ret = Browser.newContext();
```

Browser.prev()

Description:

Navigate to the previous card.

Syntax:

```
var ret = Browser.prev();
```

Parameters:

```
void
```

Returns:

An empty string.

Example:

```
var ret = Browser.prev();
```

Browser.refresh()

Description:

Refreshes current page by pulling it from the server.

Syntax:

```
var ret = Browser.refresh();
```

Parameters:

```
void
```

Returns:

An empty string.

Example:

```
var ret = Browser.refresh();
```

Browser.setVar

Description:

Sets the value of a variable.

Syntax:

```
var isSet = Browser.setVar(varName, varValue);
```

Parameters:

❑   #1: variable name

❑   #2: value to assign to variable

Returns:

Boolean True on success; else False.

Example:

```
var isSet = Browser.setVar("password", "plugh");
```

C

Java Low Memory Manager: A Development Guide

This article by Mike Kirkup (Research In Motion) appeared in the Blackberry Developer Journal *1, no. 3 (October 2004) and is reprinted by permission subject to the legal disclaimer set forth at* www.blackberry. com/developers/journal/jan_2005/legal_disclaimer.shtml. *Research In Motion Limited (RIM) assumes no responsibility for any typographical, technical, or other inaccuracies in this document. No representations or warranties by RIM may be inferred from this document, and RIM shall not be liable for any damages or other harm attributable directly or indirectly to the information contained herein.*

The *Low Memory Manager* (LMM) is used to maintain memory resources on the BlackBerry handheld when those resources pass a low-level threshold indicating that memory resources are in scarce supply. The *LMM* will attempt to free up existing memory in an effort to provide more memory space on the handheld. All applications, including third-party solutions, should work with the *LMM* to clear up as much space as possible when the handheld is running low on memory resources.

Low Memory Manager Conditions

It is important to understand that there are three conditions that can cause the *LMM* to try to free up memory resources:

1. The amount of available flash memory on the handheld decreases below a certain threshold. The free flash threshold is actually dependent on the amount of free RAM in the system. Generally, the free flash threshold varies between 400 KB and 800 KB with a guarantee to invoke the *LMM* if free flash drops below 400 KB.

2. The number of object handles available on the handheld decreases below 1,000 handles. The number of available handles depends on the amount of flash on the handheld. For 8 MB handhelds, there are approximately 12,000 available handles. For 16 MB handhelds, there are approximately 27,000 available handles.

3. The number of reference ordinals available on the handheld decreases below a certain threshold. On current handhelds this threshold is set to 1,000 reference ordinals. The number of available reference ordinals depends on the amount of flash on the handheld. For 8 MB handhelds, there are approximately 24,000 available reference ordinals. For 16 MB handhelds, there are approximately 56,000 available reference ordinals.

Working with the Low Memory Manager

There are two stages for applications to take advantage of the *LMM*. We will discuss these two stages in detail.

1. Registering your application as a `LowMemoryListener`.

2. Handling events received by the `LowMemoryListener`.

Registering Your Application as a LowMemoryListener

In order to use the *LMM*, you need to register your application with the *LMM*. The only way to do that is to provide an implementation of the `LowMemoryListener` interface in your application. When your application is started for the first time, you want to register the listener implementation with the *LMM*. Note that you only want to register your listener once. It is very important not to register more than once because you would receive multiple calls when the *LMM* is invoked.

The `LowMemoryListener` has the following method:

```
public boolean freeStaleObject( int priority )
```

This method will be automatically invoked when the *LMM* recognizes that the system is running low on memory, or when it has been invoked directly by a call to the poll method in the `LowMemoryManager` class.

In order to implement the `freeStaleObject()` method, you need to first understand the concept of where priority comes into this discussion. There are three levels of priority: low, medium, and high. In each case, the operation of the application could differ in terms of what memory resources it gives up.

For example, when `freeStaleObject()` is invoked with low priority, the application should consider clearing up some transitory variables and anything that is currently not necessary for complete functionality, such as cached data. No extraordinary efforts should be made.

When invoked with medium priority, the application should consider cleaning up stale data such as very old emails or old calendar entries.

When invoked with high priority, the application should clean out objects in the application on a *Least Recently Used (LRU)* policy. For example, when the *LMM* is invoked on the Email application it will start to delete messages at the bottom of the list of emails since those are likely to be the least used. The application should remove all the stale objects it has in order to reduce the amount of memory consumed on the handheld.

The `freeStaleObject()` method returns a Boolean, which is used to indicate if persistent data was released during the call to `freeStaleObject()`. It returns `True` if persistent data was released.

Implementation Details

Now that we have discussed the different approaches to freeing objects and how that relates to the priority, it is important to discuss how an application should go about freeing up objects. The *LMM* provides one method that enables this exact procedure.

The application should first remove all references to an object and delete it from its data structures. On completion, the application should call `LowMemoryManager.markAsRecoverable()`, passing in the object that should be freed. This command will indicate to the underlying JVM that this object can be removed as part of the *LMM* operation.

It is important to call the `markAsRecoverable()` method because the *LMM* has a specific amount of memory it is trying to recover. It will spread this amount over all of the registered applications. If your application does free up memory resources, it is important to notify the *LMM* of this fact so that it can count that freed memory against its target amount.

The following is a sample implementation of the `freeStaleObject()` method that frees up items from different persisted vectors associated with each of the priority levels:

```java
/**
 * The implementation of the freeStaleObject method required by the
 * LowMemoryListener. This method is invoked when the LMM is running out of
 * memory related resources. The application is asked to free up resources according
 * to the priority level passed into this method.
 * @param priority the priority of the call which is either High, Medium or Low.
 * @return a boolean indicating whether or not memory was freed by this call.
 * It is VERY important that the proper value is returned from this method.
 */
public boolean freeStaleObject( int priority )
{
  boolean dataFreed = false;

  switch( priority ) {
    case LowMemoryListener.HIGH_PRIORITY:
      dataFreed = freeVector( _data._high );
      _priority = LowMemoryListener.LOW_PRIORITY;
      break;
    case LowMemoryListener.MEDIUM_PRIORITY:
      dataFreed = freeVector( _data._medium );
      _priority = LowMemoryListener.HIGH_PRIORITY;
      break;
    case LowMemoryListener.LOW_PRIORITY:
      dataFreed = freeVector( _data._low );
      _priority = LowMemoryListener.MEDIUM_PRIORITY;
      break;
  }

  if( dataFreed ) {
```

```
      _persist.commit();
    }
    return dataFreed;
}


/**
 * A private method that will free up the priority vector.
 * @param vector the vector to free up.
 * @return a boolean indicating whether any objects were freed by this method.
 */
private boolean freeVector( Vector vector )
{
    boolean dataFreed = false;
    int size = vector.size();

    for( int i = size - 1; i >= 0; i-- ) {
        Object obj = vector.elementAt( i );
        vector.removeElementAt( i );
        LowMemoryManager.markAsRecoverable( obj );
        dataFreed = true;
    }

    return dataFreed;
}
```

Writing Efficient J2ME Software

Development trends constantly change with the passage of time.

In the past, the focus was on execution speed and the reduction of system overhead.

The main focus today in non-wireless development seems to be on the ease and speed of development.

In today's wireless world, current limitations are similar to past limitations, so the primary focus remains on execution speed and the reduction of system overhead.

While some wireless device limitations may lessen over time, speed and storage concerns will likely remain of top concern because of the adoption of Java to make development easier, faster, and more full featured.

Java

In theory, Java is often considered close to perfect as far as languages go. In reality, many problems have arisen related to a common view that Java is a universal language suitable for all forms of development.

This problem can be illustrated by the following examples.

In two cases, management made a decision to directly port cash cow C and C++ products to Java. They believed that creating a multi-platform product in Java would decrease long-term development costs and time to market.

The development teams bought into the concept completely and did their best to handle the port within the allotted time frame. Unfortunately, line-in-the-sand deadlines combined with crushing workloads working in a language that did not lend itself well to the port resulted in the creation of products that did not work as planned. The original applications were fast, reliable, easy to set up, and non-consumptive, but the Java versions were not.

One of the companies made the mistake of converting their C and C++ code base directly into Java hoping that the end result would work well enough to release. It did not. The direct port resulted in a poor imitation of the original, but the company released the product anyway.

Both products failed to gain sustained acceptance after release. The development teams reacted by working harder than before to optimize the code, reduce bugs, and make adoption more palatable to the customer. Unfortunately, the damage had been done and new sales went flat.

It is possible to create full-featured applications in Java, as long as common sense is used throughout the design and development stages. It is important to understand the strengths and weaknesses of the language, the Virtual Machine (VM), and the underlying hardware and network.

BlackBerry and Java

Research In Motion (RIM) chose the least-traveled route to create their Java 2 Micro Edition (J2ME)-based BlackBerry wireless handheld. The Java Virtual Machine (JVM) is optimized for the device and operating system, with native applications optimized for speed, overhead, and reliability. RIM also created a development environment and compiler that generates optimized code for the BlackBerry and extends the language with the BlackBerry API.

This was the correct, but not obvious, decision. RIM increased their development workforce and remained focused until they felt secure that their offering was world-class.

Do You Grok?

Wireless development is different than any other form of development. The limitations are similar to those encountered years ago when microcomputers were in their infancy. There are inherent limitations on the size of persistent storage and memory, available bandwidth, battery life, processor speed, and screen size. In these early stages of life for the wireless industry, the options of swapping in a faster CPU, increasing memory and storage, installing an accelerated video card, connecting a larger monitor, or accelerating transmission rates are not possible. If you write an application that makes excessive demands on resources or causes instability, your application may not work properly, may not co-exist with other applications, and may alienate your potential client base.

The Profiler

A good way to isolate and address coding bottlenecks is to use the Profiler that comes with RIM's Java Development Environment (JDE). The Profiler will display the percentage of time spent in each area of code up to the point of execution.

❑ *Summary View* — Statistics are prepared showing the percentage of time that the JVM has spent idle, running code, and performing quick and full garbage collection.

❑ *Methods View* — A module list is displayed sorted by profiled information or the number of times each item was executed.

❑ *Source View* — Source lines are displayed to navigate through methods that call, or are called by, that method.

Overall, the Profiler is a powerful tool that will help to quickly find and cure coding bottlenecks.

Memory Leaks

The JVM relies on a garbage-collection task that locates and deletes objects that are no longer needed by an application. At each pass, the garbage collector starts at the root nodes where classes exist that persist for the duration of application's life, then traverses each referenced node. During traversal, all actively referenced nodes are tracked, and all nodes that are no longer referenced are deleted, with associated memory passed back to the JVM.

If the high-persistence class fails to clear its reference to the limited persistence class after dismissal, memory will not be reclaimed until the high-persistence class goes out of scope. Memory can leak in situations where a high-persistence class creates one, or more, limited persistence classes. For example, user-interface widgets (forms, controls, and so on) that are displayed then dismissed by users.

Common places to find the cause of memory leaks are in collection classes such as vectors, list and hash tables, especially if the class has been declared static and exists for the life of the application. Another common area is when a class is registered as an event listener, yet is never unregistered before the class goes out of scope.

RIM's JDE provides a Memory Statistics tool and an Object tool that will display statistics on the number of objects and bytes that are used for object handles, RAM, and flash memory. With memory statistics, you set a few breakpoints, run your application, refresh and retain the memory statistics, run the application to the next breakpoint, and then refresh to compare memory statistics to the previous breakpoint.

Bandwidth and Latency

Java-based BlackBerry now supports operation on three different types of wireless networks: GPRS, CDMA (1X), and iDEN. The bandwidth on any given network can fluctuate from less than 5 kilobits/second (kbps) to as high as 50 kbps. This variance in bandwidth would generally be acceptable if it were not for network latency. First round-trip latency usually takes 5 seconds or more (depending on network traffic) to complete, while consecutive round-trip attempts take, on average, 2 seconds or more.

Wireless applications should compensate by minimizing the amount of data transferred over the network, and hide latency delays from users. Utilizing the multiple-thread architecture of the BlackBerry is an effective way to improve both perceived and actual performance by handling more than a single task at the same time. For example, BlackBerry email is compressed, then transmitted as a background thread to allow use of the unit while network activity is present. This provides the user with the perception of speed. Conversely, a Web browser user must wait for a page to come up and does not expect to do anything else while waiting. A well-designed wireless Web site will reduce the impact of latency by minimizing the number of round trips a browser needs to fetch a page.

General Guidelines

The following are general guidelines that may be useful in day-to-day development:

❑ Create a new thread for any lengthy operations such as network connections. Use background threads for listeners or other processes that you want to run in the background when the application starts.

❑ When you use an inner class to hide one class inside another, but you do not need the inner class to reference the outer class object, declare the inner class static. This suppresses the creation of the reference to the outer class.

The only time you should use a non-static inner class is when you need access to data in the outer class from within methods of the inner class. If you are using an inner class only for name scoping, make the inner class static.

❑ When creating code libraries, ensure that you mark a class as final if you know it will never be extended. The presence of the final keyword enables the compiler to generate faster code.

❑ In Java, a long is a 64-bit integer. Since the BlackBerry wireless handheld uses a 32-bit processor, operations run two to four times faster if you use an int instead of a long.

❑ When defining static fields (also called *class fields*) of type String, you can increase program speed by using static variables (not final) instead of constants (final). The opposite is true for primitive data types, such as int.

For example, you might create a String object:

```
private static final String x = "example";
```

For this static constant (denoted by the `final` keyword), a temporary String instance is created every time you use the constant. The compiler eliminates x and replaces it with the string `"example"` in the bytecode, so that the virtual machine (VM) has to perform a hash table lookup each time you reference x. In contrast, for a `static` variable (no final keyword), the String is created once. The VM performs the hash table lookup only when it initializes x, which makes access faster.

The BlackBerry JDE compiler automatically marks any classes as final that are not extended in an application .cod file.

It is acceptable to have public constants (that is, final fields), but you should always make variables private.

❑ Use local variables wherever possible because access to local members is far more efficient than access to class members.

❑ Interfaces produce larger, slower code. Avoid creating interfaces when creating API libraries unless you foresee multiple implementations of the API.

❑ Use appropriate access modifiers for fields and methods when creating code libraries to reduce the size of the compiled code. Declare fields as private whenever possible, and use the default (package) access instead of public access (that is, public and protected keywords).

❑ Use or extend a class from the J2ME and BlackBerry APIs where possible, and use inheritance at every turn.

❑ To increase speed, either inline the code in critical sections or reduce size by placing duplicate code in its own method.

❑ Avoid using a lot of date objects to save space. Convert date values into longs.

❑ Synchronized methods are twice as expensive as regular methods. If necessary, synchronize on methods rather than code blocks to slightly improve performance.

❑ Try/catch exception handling is handled efficiently by the BlackBerry VM and does not introduce expensive overhead like other Java applications.

❑ Avoid recursive calls in critical sections of code to increase speed. Use recursion in less-critical sections to reduce code size.

❑ Use native classes for machine performance that can't be matched in Java. For example, `arraycopy()` is considerably faster than copying a large array in a loop.

❑ Minimal savings can be found by using short names for file paths, classes, methods, and instance variables. Unlike .JAR format, applications created in .COD format are tightly compressed and obfuscated.

❑ Some API classes are so full-featured that you'll take a speed and size hit when you only have to use a subset of the functionality. If you run into this situation, profile your code using the API class, substitute your own simplified method, then profile again. The difference can be substantial in high traffic areas of code.

❑ If you encounter performance issues with a method within an API class, override the method with a simplified, optimized version of your own.

❑ Replace expensive Java data structures with simpler data structures to increase speed at a cost of code complexity. For example, multi-dimension arrays are, by far, more expensive than single-dimension arrays because of the extra indirection.

❑ Switches use fast or slow search algorithms depending on how close the switched values are. A (fast) direct lookup is used for close values where a (slow) table search is used for disparate values.

❑ Methods can only be inlined when they are final, private, or static, and lack local variables. If possible, rewrite high-traffic methods to be final sans local variables for inlining.

❑ Reuse objects wherever possible. It's expensive to create a new object and eventually leads to garbage collection. In comparison, `'Object obj = new Object()'` is 43 times more expensive than declaring `'int i=0'`. To put the BlackBerry VM in a more favorable light, it is upward of 3,000 times more expensive to create an object within a conventional applet than `'int i=0'`.

❑ Avoid using immutable objects to save on garbage collection from discarded objects left behind after the creation of a new, altered immutable objects.

The following code snippet demonstrates how to create garbage with an immutable String:

```
public String FibonacciString1(int start, int end)
{
  String s = new String();

  for (int i = start; i <= end; i++){
    s = s + Fibonacci1(i);
    s = s + " ";
  }
  return s;
}
```

An immutable String called 's' is created outside of the loop. Within the loop, 's' is discarded and created, then discarded and created again, as many times as necessary thus leaving garbage on the heap at each pass. A less-consumptive approach would be to create a mutable StringBuffer then append data into the buffer:

```
public String FibonacciString2 (int start, int end) {
  StringBuffer s = new StringBuffer();

  for (int i = start; i <= end; i++){
    s.append(Fibonacci2(i));
    s.append(" ");
  }
  return s.toString();
}
```

❑   If your application relies on images that are in Portable Network Graphics (.PNG) format, make sure to export them in the most compressed form possible in the same format. Some imaging packages save .PNG images in a format that can be considerably reduced through supported compression techniques. As an example, I created a 32 by 32 bit, 24-bit color image using such a product. The size was 29,582 bytes in original form, and 1,050 bytes in compressed form.

❑   Reduce the number of custom classes in your application as much as possible. Each class definition increases the application size. Combine related methods beneath a single class to conserve space.

❑   Use clipping to reduce the amount of work done in repaint().

❑   Use high-level graphic primitives where possible to avoid repeat calls to lower-level primitives.

In Closing . . .

Take pride in creating small, fast, and bug-free wireless applications. Oppose unrealistic deadlines at every turn. Slow, bloated, and buggy applications don't make the grade in the wireless world.

Enjoy the wireless experience!

User Interface Coding Tips

This article by Scott Wahl (Manager, Software Documentation, RIM) appeared in the Blackberry Developer Journal, 1, *no. 1 (November 2003) and is reprinted by permission subject to the legal disclaimer set forth at* `www.blackberry.com/developers/journal/jan_2005/legal_disclaimer` `.shtml.` *Research In Motion Limited (RIM) assumes no responsibility for any typographical, technical, or other inaccuracies in this document. No representations or warranties by RIM may be inferred from this document, and RIM shall not be liable for any damages or other harm attributable directly or indirectly to the information contained herein.*

Three Tips for Managing Screens

1. Pop screens off the user interface (UI) stack when the user finishes interacting with them to avoid memory problems as the display stack grows.

2. Never pop the same screen more than once. Excess popping can lead to problems with the keyboard and thumb wheel.

3. Avoid using more than a few modal screens at one time because each screen uses a thread.

Handling Threads

An application can access the UI either on the event thread or with the event lock held. Only one thread at a time (usually the event-dispatching thread) can gain access to an interface component. There are two ways for background threads to access the UI from outside the main event-handling or UI drawing code:

1. Acquire and hold the event lock.

2. Use `invokeLater()` or `invokeAndWait()` to run on the event dispatch thread.

Holding the Event Lock

The event dispatcher sets a lock on the event thread while it processes a message. Background threads can access the UI by acquiring this lock for a short time without interfering with processing by the event dispatcher. A worker thread can call `Application.getEventLock()` to retrieve the event lock and then synchronize this object to ensure serialized access to the UI. You should only hold this lock for short periods of time because the event dispatcher is paused. An application should never call `notify()` or `wait()` on this object.

For example, in a timer task, you could write code in the following method:

```
class MyTimerTask extends TimerTask {
  public void run() {
    synchronized(Application.getEventLock()) {
      _label.setText("new text " + System.currentTimeMillis());
    }
  }
}
```

In most cases, this is the most efficient way to access the UI.

Running on the Event Dispatch Thread

In some cases, holding the event lock is not appropriate, particularly if the application has its own locking operations to manage. In this case, you must create a class that implements the `Runnable` interface. You can then invoke its `run()` method on the event dispatch thread using one of these three methods:

1. `invokeAndWait(Runnable runnable)`

Immediately calls `run()` on the event dispatch thread. The call blocks until the `run()` method is completed.

2. `invokeLater(Runnable runnable)`

Calls `run()` on the event dispatch thread after all pending events are processed.

3. `InvokeLater(Runnable runnable, long time, Boolean repeat)`

Calls `run()` on the event dispatch thread after a specified amount of time, where time specifies the number of milliseconds to wait before adding `Runnable` to the event queue. If repeat is true, `Runnable` is added to the event queue every `time` milliseconds.

Using activate()

The system calls the `activate()` method when it brings an application to the foreground. For most applications, you should not override `activate()`. Perform any initialization, including any required `pushScreen()` calls, in the application's constructor. If you must override `activate()`, ensure that you call `super.activate()` from within the overridden method; otherwise, your application does not repaint correctly. Because `activate()` can be called multiple times for the same application, it is not appropriate to perform a one-time initialization with this method. For example, you should not perform actions, such as call `pushScreen()` from within `activate()`, unless you verify that you have done so already.

Exiting Applications

When the user closes an application (for example, by selecting the Close menu option), most applications should shut down using `System.exit()` before returning the user to the Home screen. This conserves space on the handheld.

Storing Data Persistently

Persistently trying to resolve issues, persisting to an old age, and the persistent odor of a light perfume. These can all be good things. Committing facts to memory, writing a shopping list, taking notes in class, sending email, saving documents to disk, and storing information within a database are all good examples of persistence.

In the wireless world of BlackBerry development, persistence can be found when sending and receiving email; making browser requests; pushing and pulling data; syncing and adjusting calendar, task list, contacts, and notepad data; installing applications over the air; and storing data programmatically.

Here we will address the programmatic storage of data on the BlackBerry handheld.

Considering the Facts

Creating an application for the BlackBerry handheld requires a developer to consider many factors not present in a conventional system, such as:

- ❑ Persistent store objects live in a finite amount of flash memory that is shared with the operating system, BlackBerry applications, third-party applications, and user data.

- ❑ The data transfer rates to and from flash memory is slower than conventional memory (RAM).

- ❑ Data transmission speeds across wireless networks are far lower than across wireline networks.

❑ Reading and writing to flash memory forces the CPU to work harder, which in turn compromises battery life.

❑ Physical representations of persistent data are restricted to byte arrays for MIDlets (and Java objects when writing specifically for the BlackBerry API).

Two Ways to Do It

There are two ways to programmatically store data on a Java-based BlackBerry handheld:

1. MIDP record stores
2. The BlackBerry Persistence API set

MIDP record stores should be used when writing applications that will work on all J2ME-enabled devices. MIDP record stores have a lot of methods available to manipulate them, and are relatively easy to work with.

The most useful feature is the ability to read, insert, delete, and update individual records within a store of records without having to load the entire store into memory first. The down side is that you can only store byte arrays, cannot retain more than 64 KB of data within a record store, and automatically provide data visibility to MIDlets within the same MIDlet suite.

The signed BlackBerry persistence model can be used to develop applications that take full advantage of the BlackBerry architecture. Size restrictions are largely removed, data visibility is set by the application, and objects (including custom objects) can be saved to persistent store. Persistently stored data can also be backed up and restored through the use of the signed synchronization API in the `net.rim.device`
`.api.synchronization` package.

Note that `PersistentStore` is a lightweight database solution. You can serialize an object of any type to persistent store, but cannot selectively update or locate elements. You must load the entire object into memory, alter as required, and then commit the entire object back to persistent store.

Code Signing

The BlackBerry persistence model requires the use of signed APIs. For more information, please review Jonathan Nobels's article, "Give Me A Sign," which has been published in the "BlackBerry Developer Journal," vol. 1, no. 2 (May 2004).

BlackBerry PersistentStore and PersistentObject

Serialization of data is straightforward with the BlackBerry API.

Create/Open

Call `PersistentStore.getPersistentObject(long_key)` to retrieve a reference to an existing `PersistentObject`, or to create a new one if it does not exist. Use a static constructor so that only one `PersistentObject` is created the first time that an object of this class is created. Each time a process starts, the static block will be run again.

```
public class MyStore implements KeyListener, TrackwheelListener {
  private static PersistentObject store;
  static {
    // key hash of com.rim.bbdj.mystore
    store = PersistentStore.getPersistentObject(0xa406067aeb8ca6ebL);
  }
```

Note that `long_ key` must be a unique long value. The easiest way to create a unique value is to create a hash of your fully qualified package name. If you use more than one persistent store object within your application, append a descriptive table identifier to your package name before deriving the hash. The steps required to create a hash value from within the BlackBerry IDE are:

1. Type a string value, such as `com.rim.bbdj.mystore`

2. Highlight the entire string

3. Right-click then click "Convert 'com.rim.bbdj.mystore' to long"

The long value appears (`0xa406067aeb8ca6ebL`). Make sure to include a comment in your code to indicate the string used to generate the long key.

Store

Use `PersistentObject.setContents(Object obj)` then `PersistentObject.commit()` within a synchronized block to store data.

```
private MenuItem saveItem = new MenuItem("Save", 110, 10) {
  public void run() {
    String username = new String("Tiberius");
    String password = new String("IGrokSpock");
    String[] userinfo = {username, password};

    synchronized(store) {
      store.setContents(userinfo);
      store.commit();
    }
  }
};
```

Retrieve

Use `PersistentObject.getContents()` within a synchronized block to retrieve data.

```
private MenuItem getItem = new MenuItem("GetItem", 110, 11) {
  public void run() {
    synchronized(store) {
      if(store.getContents() == null)
        System.out.println("Error");
      else
        String[] currentinfo = (String[])store.getContents();
    }
  }
};
```

Delete

Use `PersistentStore.destroyPersistentObject( long id )` to delete the database.

```
PersistentStore.destroyPersistentObject(0xa406067aeb8ca6ebL);
```

Custom Persistent Objects

Custom objects can be stored persistently. The process is basically the same when creating/opening, storing, retrieving, and deleting a persistent store object.

There are two main differences. First, you need to work with anything that extends `Object`. Second, the class of the object to be saved must implement the `Persistable` interface.

The main differences in sequence:

1. Create a `Vector` object to store multiple objects

```
private static Vector data;
```

2. Create a `PersistentObject` database

```
private static PersistentObject store;
```

3. Initialize the database to store a `Vector`

```
static {
  store = PersistentStore. getPersistentObject(0xa406067aeb8ca6ebL);

  synchronized (store) {
    if (store.getContents() == null) {
      store.setContents(new Vector());
      store.commit();
    }
  }
  data = new Vector();
  data = (Vector)store.getContents();
}
```

4. Create a `Persistable` class to handle the `Vector` data

```
private static final class RestaurantInfo implements Persistable {
  //data
  private Vector elements;

  //fields
  public static final int NAME = 0;
  public static final int ADDRESS = 1;
  public static final int PHONE  = 2;
  public static final int SPECIALTY = 3;

  public RestaurantInfo() {
    elements = new Vector(4);

    for ( int i=0; i<elements.capacity(); ++i)
```

```
        elements.addElement(new String(""));
    }

    public String getElement(int id) {
      return (String)elements.elementAt(id);
    }

    public void setElement(int id, String value) {
      elements.setElementAt(value, id);
    }
}
```

MIDP Record Stores

MIDP Record Management System (RMS) provides developers with the ability to create persistent store databases, then selectively add, update, and delete individual rows of byte array data within the database.

Multiple MIDlets within a MIDlet suite, or threads within a MIDlet, can open databases, and read/write records within a persistent store database. This can be performed without clash because all record store operations are atomic, synchronous, and serialized. To ensure data integrity, locking of the entire `RecordStore` is employed when reading/writing individual records within the store. While table locking is unacceptable with conventional relational database systems, it was deemed acceptable on wireless devices as the overhead for memory/storage and processing to maintain row-level locks would be far too excessive.

MIDP relies on the `javax.microedition.rms.RecordStore` class and four interfaces for most operations:

1. `RecordComparator`
2. `RecordEnumerator`
3. `RecordFilter`
4. `RecordListener`

Each MIDlet suite has its own separate name space for record stores. This means that MIDlets can access any record store within their MIDlet suite.

Create/Open

Create a new `RecordStore` database by calling `RecordStore.openRecordStore()`, passing the name of the database, and a Boolean flag set to `'true'` to create the database if it doesn't exist.

```
String name = "myDb";
RecordStore rs = RecordStore.openRecordStore( name, true );
```

Best practice is to iterate through all `RecordStores` in the MIDlet suite to determine if a naming clash exists before creating or opening a database. This can be accomplished as follows:

```
String [] stores = RecordStore.listRecordStores();

if ( stores!= null ) {
   for ( int element = 0; element < stores.length; element++ ) {
      if ( stores[element].equals(name)) {
         // name is already in use
      }
   }
}
```

It is best to check before creating a new database for the following reasons:

❑ Any MIDlet within a MIDlet suite can create databases that are shared within the suite

❑ RecordStore names are limited to 32 Unicode case-sensitive characters

❑ No guidelines are in place to construct filenames

Store

There are two ways to store record data.

The first method adds a new record to a RecordStore database:

```
// addRecord(byte[], byte[]);
int recordId = rs.addRecord(data, 0, data.length);
```

Pass a byte [] buffer populated with the data to write, the offset into the buffer where the data starts, and the number of bytes to write. The example uses data.length, which assumes that the full buffer is being used. This method returns the sequential record identifier assigned to the new record.

The second method updates an existing record within a RecordStore database:

```
// setRecord( int, byte[], int, int );
rs.setRecord(recordId, data, 0, data.length);
```

Pass the record identifier, a byte [] buffer populated with the data to write, the offset into the buffer where the data starts, and the number of bytes to write.

Retrieve

There are two ways to retrieve a RecordStore record.

The easiest is to simply pass the record identifier you want:

```
// getRecord( int );
byte [] data = rs.getRecord(recordId);
```

The second method is a little more powerful because it allows you to selectively append data to the byte [] buffer:

```
// int getRecord( int, byte[], int );
int bytes_copied = getRecord(recordId, data, offset);
```

Pass the record identifier, a buffer to hold the `byte []` data, and the starting offset into the buffer to write data. The number of bytes retrieved into the passed `byte []` buffer (data) is returned.

Delete

Delete an individual record via:

```
// deleteRecord( int );
rs.deleteRecord(recordId);
```

Delete the entire `RecordStore` database via:

```
// deleteRecordStore( String );
rs.deleteRecordStore(recordStoreName);
```

Close

```
if ( rs != null )
  rs.closeRecordStore();
```

Row Iteration

`RecordStore.enumerateRecords()` is used to step through the rows within an active `RecordStore` database. The returned `RecordEnumeration` object can be used to work forward or backward through a record set as defined when first calling `enumerateRecords`. This process can be made to simply walk through all rows without regard to sequence, or can be enhanced to track row changes, filter results, or custom-sort underlying rows.

Methods available within the `RecordEnumeration` interface include:

❑ `public byte[] nextRecord();`

❑ `public byte[] previousRecord();`

❑ `public int nextRecordId();`

❑ `public int previousRecordId();`

❑ `public boolean hasNextElement();`

❑ `public boolean hasPreviousElement();`

❑ `public int numRecords();`

❑ `public void keepUpdated( boolean keepUpdated );`

❑ `public boolean isKeptUpdated();`

❑ `public void reset();`

❑ `public void rebuild();`

❑ `public void destroy();`

The following code can be used to access all rows within a RecordStore database in sequence. Note that the parameters passed to enumerateRecords() disable RecordFilter and RecordComparator, and also disable automatic updates to the underlying rows if any external changes occur during processing.

```
// open, but not create, database
RecordStore rs = RecordStore.openRecordStore( "myStoreDB", false );

if (rs != null ) {
  RecordEnumeration dataset = null;

  try {
    // parm 1: RecordFilter [ disabled ]
    // parm 2: RecordComparator [ disabled ]
    // parm 3: track changes [ disabled ]
    // RecordEnumeration enumerateRecords
    // (RecordFilter, RecordComparator,
    //  boolean);
    dataset = rs.enumerateRecords( null, null, false );

    while(dataset.hasMoreElements()) {
      int recordId =dataset.getNextRecordId();

      // row identifier is valid
      if ( recordId )
        byte [] data = rs.getRecord(recordId);
    }
  }
  catch( RecordStoreException except ) {
  }
  finally {
    dataset.destroy();
  }
  rs.closeRecordStore();
}
```

RecordEnumeration sets can be traversed forward and backward.

For forward navigation, use RecordEnumeration.getNextRecordId() in combination with RecordStore.getNextRecord() to get the next row, and RecordEnumeration.hasNextElement() to determine if additional rows are available.

For backward navigation, use RecordEnumeration.getPreviousRecordId and RecordStore.getPreviousRecord() to get the previous row, and RecordEnumeration.hasPreviousElement() to determine if additional rows are available.

Call RecordEnumeration.reset() at any time to change the direction, and reset access at the start.

RecordStore.enumerateRecords Parameters

Parameter 1: RecordFilter

Passing a RecordFilter object to enumerateRecords() will pass the byte [] data within all processed rows to the matches() method of the RecordFilter object before allowing processing to occur.

The `RecordFilter` interface contains a single method as shown here:

```
public interface RecordFilter {
  public boolean matches( byte[] recordData );
}
```

A sample implementation could work as follows:

```
dataset = rs.enumerateRecords( new DSFilter(), null, false );

...

public class DSFilter implements RecordFilter{
  public boolean matches( byte[] recordData ){
    boolean ret = false;

    if (recordData[0] != 0 && recordData.length > 0)
      ret = true;

    return ret;
  }
}
```

Parameter 2: RecordComparator

Pass a `RecordComparator` object to `enumerateRecords()` to sort rows. It contains a single method as shown here:

```
public interface RecordComparator {
  public int compare(byte[] rec1,byte[] rec2);

  public int EQUIVALENT =  0;
  public int FOLLOWS    =  1;
  public int PRECEDES   = -1;
}
```

A sample implementation could work as follows:

```
dataset = rs.enumerateRecords( null, new DSCompare(),false );

...

public class DSCompare implements RecordComparator {
  public int compare(byte[] rec1, byte[] rec2){
    String srec1 = new String(rec1);
    String srec2 = new String(rec2);

    return (srec1.compareTo(srec2));
  }
}
```

Note that there will be a speed penalty when using RecordComparator, especially when the third parameter, keepUpdate, is set to true. In this situation, the entire data set will automatically be resorted after the RecordEnumeration index has been rebuilt.

Parameter 3: keepUpdated

The third parameter to enumerateRecords() should be set to 'true' only if there is a risk that some rows will be externally changed, added, or deleted during navigation. In this situation, the RecordEnumeration will become a listener of the RecordStore and react to record additions and deletions by re-creating its internal index. Do not use this feature unless absolutely warranted, as there will be a performance hit every time the index is rebuilt due to a change to the RecordStore.

Summary

If you do not require the cross-platform capability offered by MIDP record stores, use the BlackBerry persistence API set to take full advantage of the BlackBerry handheld architecture.

Index

Symbols and Numerics